New Approaches for Multi-Criteria Analysis in Building Constructions

Valentino Sangiorgio · Luis G. Vargas ·
Fabio Fatiguso · Francesco Fiorito

New Approaches for Multi-Criteria Analysis in Building Constructions

User-Reporting and Augmented Reality to Support the Investigation

Springer

Valentino Sangiorgio
ICITECH Department
Polytechnic University of Valencia
Valencia, Spain

DICATECh Department
Polytechnic University of Bari, Bari, Italy

FEUP Department
University of Porto
Porto, Portugal

Fabio Fatiguso
DICATECh Department
Polytechnic University of Bari
Bari, Italy

Luis G. Vargas
The Joseph M. Katz Graduate School
of Business
International Center for Conflict Resolution
University of Pittsburgh
Pittsburgh, PA, USA

Francesco Fiorito
DICATECh Department
Polytechnic University of Bari
Bari, Italy

ISBN 978-3-030-83874-4 ISBN 978-3-030-83875-1 (eBook)
https://doi.org/10.1007/978-3-030-83875-1

This Springer imprint is published by the registered company Springer Nature Switzerland AG
The registered company address is: Gewerbestrasse 11, 6330 Cham, Switzerland

Authors' Contribution

This book is written by *Sangiorgio Valentino, Vargas Luis G., Fatiguso Fabio* and *Fiorito Francesco.*

Sangiorgio Valentino (corresponding Author) dealt with conceptualization, writing (original draft preparation), data curation, formal analysis, investigation and visualization. *Vargas Luis G., Fatiguso Fabio* and *Fiorito Francesco* dealt with review, editing and supervision.

In particular, the contribution of each author is reported for every chapter of the book in the following.

Chapter 1: Introduction—*Sangiorgio Valentino, Vargas Luis G., Fatiguso Fabio* and *Fiorito Francesco.*
Chapter 2: The Analytic Hierarchy Process in the Building Sector—*Sangiorgio Valentino* and *Vargas Luis G.*
Chapter 3: Augmented Reality to Support the Analytic Hierarchy Process—*Sangiorgio Valentino, Vargas Luis G.* and *Fatiguso Fabio.*
Chapter 4: How to Set a User Reporting Supported Decision Making in Architectural Engineering and Building Production—*Sangiorgio Valentino, Vargas Luis G., Fatiguso Fabio* and *Fiorito Francesco.*
Chapter 5: AR-AHP to Support the Building Retrofitting: Selection of the Best Precast Concrete Panel Cladding—*Sangiorgio Valentino, Vargas Luis G.* and *Fatiguso Fabio.*
Chapter 6: User Reporting and AHP to Investigate the Perception and Social Acceptance of Wind Energy—*Sangiorgio Valentino, Vargas Luis G.* and *Fiorito Francesco.*
Chapter 7: User Reporting and Condition Ratings to Support Building Maintenance and Diagnostics—*Sangiorgio Valentino, Vargas Luis G.* and *Fiorito Francesco.*

About This Book

Multi-criteria decision analysis (MCDA) is widely used in many sectors and increasingly in the field of building constructions due to their flexibility and adaptability in different areas. Indeed, in recent years, different stakeholders exploited MCDA to support all the different phases of the **building process**: *design*, *construction*, *management* and *dismantling*.

This book aims at supporting researchers, practitioners, architects, engineers and students to use **interdisciplinary multi-criteria decision analysis** in the **building construction sector with a specific focus on architectural engineering and building production**. In the context of the "digital transition" of the building sector, this book focuses on the use of innovative approaches such as "**User Reporting**" and **augmented reality** supporting the **multi-criteria decision analysis**. In addition, the proposed methods allow including both quantitative and qualitative information (such as **users' perception**) in the analysis.

Firstly, the book explains some classical multi-criteria decision analysis methods such as the **analytic hierarchy process** and the **Simos-Roy-Figueira method**, to provide the reader an advanced understanding of MCDA by explaining every single phase of the process. The use of MCDA is clarified for different purposes, e.g. achieving concise performance indexes, identifying the best/most preferred solution among a set of *alternatives* or investigating stakeholder's *perceptions*.

Secondly, the new approach named **augmented reality-AHP** (AR-AHP) is proposed to provide a modern and useful tool to examine the parameters involved and the possible *alternatives*. AR-AHP exploits an **augmented reality environment** combined with the AHP and an improved Simos-Roy-Figueira method to provide a large amount of visual information during the decision phase. Thanks to the use of 3D models in AR, this approach is particularly effective in the field of building construction where technical drawing, axonometry representation and three-dimensional models have been used for centuries for their ability to visually transmit a useful information to the user.

Thirdly, theory and application of "**User Reporting**" in building constructions are considered. This approach in combination with modern technologies offers an unprecedented data acquisition capacity at the scale of the individual. Indeed, the reader is guided in the following eight steps useful to obtain an effective "User Reporting" process: (i) definition of stakeholders; (ii) definition of users; (iii) selection of technological tools to support the acquisition (e.g. smart devices); (iv) creation of specific questionnaires and choice experiments; (v) definition of the flow chart describing the acquisition process; (vi) registry data acquisition; (vii) data processing; and (viii) data analysis.

All the educational contents are supported by adequate images with effective and **modern graphics** and **simplified examples**.

Finally, to provide some **examples of exceptional academic value**, the book presents some complex problems in the field of **building construction technologies**, faced by applying the proposed MCDA techniques emphasizing their synergies and potentials with novel technologies. In particular, all the different purposes of the MCDA approach discussed in the theoretical pat of the book are explained from a practical point of view exploiting specific case studies.

Contents

Abbreviations

APP	Application software
AR	Augmented reality
BCr	Building condition rating
CCr	Component condition rating
CI	Consistency index
Cr	Condition rating
CR	Consistency ratio
DM	Decision maker
MCDA	Multi-criteria decision analysis
MCDM	Multi-criteria decision method
QDP	Quality detection platform
RC	Reinforced concrete
RI	Random consistency index

Parameters: General term for *criteria, sub-criteria* or *alternatives.*

Chapter 1
Introduction

Abstract Multi-criteria decision analysis (MCDA) is widely used in many sectors and increasingly in the field of **architectural engineering** and **building production** due to their flexibility and adaptability in different areas. Indeed, in recent years, different stakeholders exploited MCDA to support all the different phases of the **building process:** *design*, *construction*, *management* and *dismantling*. In this chapter, **firstly** a general introduction of the MCA is proposed by explaining the advantages of these approaches and providing an overview of the most used methods. **Secondly**, the focus is directed on the use of MCA in the building construction sector by proposing a classification of the goal typologies that can be reached in this field: (A) obtain *Key Performance Indicators* (KPIs), specific *Condition Ratings* (*Cr*) or risk indexes to identify the conditions, performances or level of risk of the investigated buildings respectively; (B) Identify the best solution among a set of *alternatives*; (C) investigate the stakeholder's *perception*. **Thirdly** the most used multi-criteria approaches are introduced and the new related perspective concerning the synergy with modern technologies, such as Augmented Reality and User Reporting, are emphasized.

Keywords Multi-criteria decision analysis · Architectural engineering · Building production · Key performance indicators · Condition Ratings · Risk index · Best alternative · Stakeholder's perception

Multi-criteria decision analysis (MCDA) is widely used in many sectors and increasingly in the field of **architectural engineering** and **building production** thanks to the flexibility and adaptability of these approaches. In this chapter, we first introduce the MCDA proposed by explaining the general advantages of the approach and providing an overview of the most used methods. Second, we direct the focus to the use of MCDA in the building construction sector by proposing a classification of the typologies of the goals that can be achieved in this field. Third, we describe one of the most used approaches, the *Analytic Hierarchy Process* (AHP) and discuss the new perspectives of the use of MCDA in combination with modern technologies such as Augmented Reality and User Reporting.

V. Sangiorgio et al., *New Approaches for Multi-Criteria Analysis in Building Constructions*, https://doi.org/10.1007/978-3-030-83875-1_1

1.1 Why You Should Be Interested in Multi-criteria Analysis?

The problems that could arise during the **building process** (design, construction, management and dismantling) are typically characterized by multiple actors with often conflicting values and views, a wealth of possible outcomes and high uncertainty. Often, it is of basic importance to consider different aspects and heterogeneous measures, including both *qualitative* and *quantitative* parameters in the analysis, to face these problems.

This necessity leads to a question: *Are there mathematical and engineering approaches to consider parameters of different typologies in order to choose the best solution?*

To provide some examples, at the *urban scale* it could be difficult to decide the best urban renovation project among offices, universities, academies, residential areas, green and park and so on (Fig. 1.1a). On the other hand, even at the *scale of the individual construction*, it could be necessary to identify the best *solution* among different alternatives for the retrofitting and renovation intervention (Fig. 1.1b).

Whatever the scale of application, the identification of the best *solution* among a set of different alternative projects (or among different management, retrofitting, or renovation strategies) is a complex decision problem in which different aspects need to be simultaneously considered. Both technical and *quantitative* criteria, which are based on numerical observations and non-technical and *qualitative* criteria, which are based on social visions, preferences and feelings need to be considered. For instance, *quantitative* criteria are the parameters directly measurable such as time, costs, environmental impact, construction performance and so on; *qualitative* criteria represent dimensionless parameters such as stakeholders needs, or some type of risks. A generic illustration of different typologies of *criteria* that can be involved in a decision problem are shown in Fig. 1.2. For example, in the design phase of a new building, criteria such as time, costs, performances or environmental impact

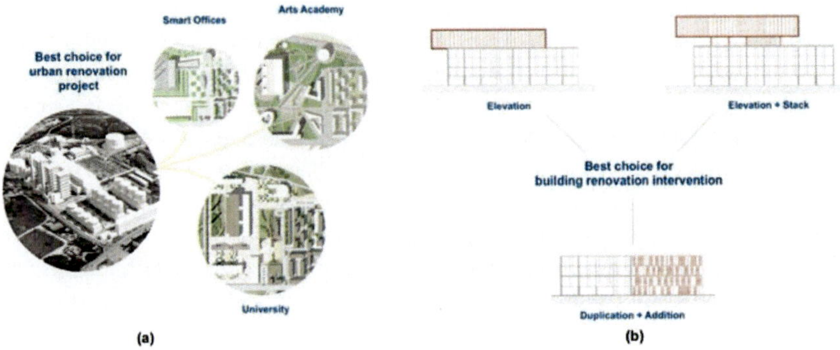

Fig. 1.1 Examples of two classical decision problems in the construction sector: **a** at the *urban scale,* and **b** at the *scale of the individual construction*

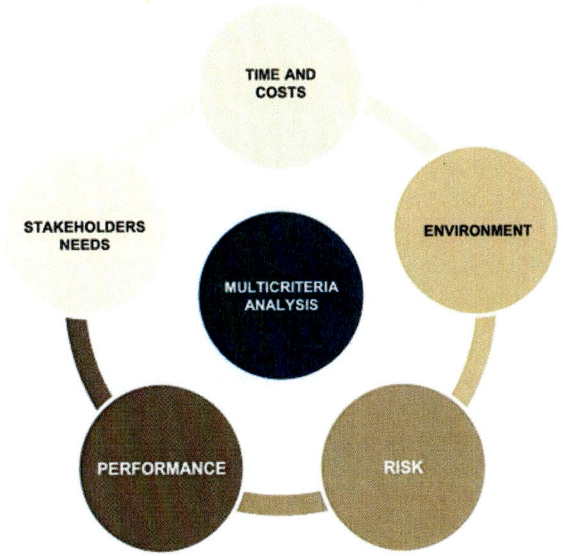

Fig. 1.2 Generic example of different *criteria* in a decision problem in the building construction sector

are fundamental for the feasibility of the project, but also qualitative data related to the risk or stakeholders' needs are important for the effectiveness of the future construction. Multiple criteria approaches are useful to analyze all these parameters together.

In this context, a very useful aid is provided by the techniques that are part of the family of Multiple Criteria Decision Analysis (MCDA), that are used to make a comparative assessment of different *solutions* by considering heterogeneous measures.

1.2 Introduction to Multi-criteria Decision Analysis

In MCDA, several *criteria* can be taken into account simultaneously in a complex situation to identify the best solution among a set of *alternatives*. The methods are designed to help a *Decision Maker* (DM) to integrate the different options, reflecting the opinions of the actors concerned, into a prospective or retrospective framework (Figueira et al., 2005). Participation of the DM in the process is a central part of the approach and the results are usually directed at providing operational advice or recommendations for future activities.

In addition, MCDA can establish preferences among *alternatives* (different *solutions*) by reference to an explicit set of objectives that the decision making body has identified and for which it has established measurable *criteria* to assess the extent to which the objectives can be achieved.

An important initial consideration in the choice of MCA technique regards the number of *alternatives* to be appraised. Some problems are concerned with outcomes that are "infinitely" variable (Azzabi et al., 2020). However, most building design and engineering decisions, even at fairly low levels, are usually about choices between discrete options, for example, between alternative design projects, or alternative intervention strategies, material and supplier.

The MCDA approaches proposed in this book are concerned with techniques for handling choices among a finite number of options. Solving problems involving optimizing continuous variables requires quite different procedures.

In general, the MCDA is achieved by developing the following points:

1. Establishing the decision context and specify the *aims* of the problem, who are the *decision makers* and other key players;
2. Identifying the *alternatives* involved;
3. Identifying the *goal* and related *macro-criteria*, *criteria* and *sub-criteria* that reflect the value associated with the consequences of each *alternative*;
4. Describing the expected performance of each *alternative* against the *macro-criteria*, *criteria* and *sub-criteria*;
5. Assigning *local weights* (weighting) for each of the *macro-criteria*, *criteria* and *sub-criteria* to reflect their relative importance in the decision;
6. Combining the *local weights* for each of *alternative* to derive *global weights* and obtain the *final ranking* of the *alternatives*;
7. Examining the results; and
8. Conducting a *sensitivity analysis* of the results to changes in *local weights* to understand if the *final ranking* is robust.

Note that these points represent the conventional procedure for some typical problems in the engineering field. On the other hand, for more complex problems, the procedure can be modified and improved until a different and more effective approach is obtained. To provide an example, some *calibration* and *validation* procedures could be included to combine the expertise of the DMs with sophisticated data analysis.

1.3 Overview of Classical Multi-criteria Approaches

There are several MCDA approaches in the technical and scientific world that can be principally classified into four categories, *empirical*, *numerical*, *monetary* and *non-monetary*.

The *empirical* analyses are not based on the determination of weights and do not use mathematical approaches (algorithms or equations) as in the case of *numerical*

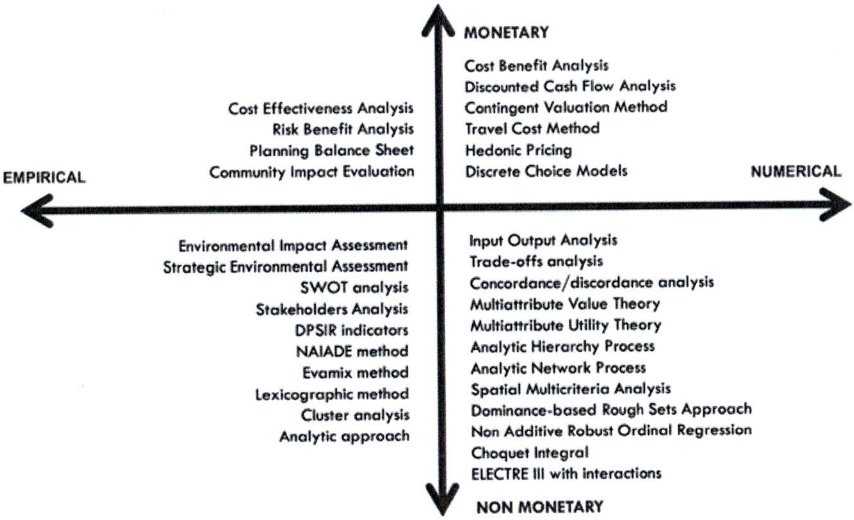

Fig. 1.3 Classification of common multi-criteria evaluation methods

ones. On the contrary, *empirical* analyses are based on strategic and methodological planning techniques to address the problem in a qualitative way.

The *monetary* analyses are aimed at the economic evaluation of the multi-criteria problem.

The *non-monetary* analyses are more flexible and can consider both quantitative and qualitative parameters in the evaluation including, in some cases, also the economic evaluation among the involved parameters.

The most common MCA approaches are listed and classified according to the defined four categories in Fig. 1.3.

This book is focused on *numerical* and *non-monetary* analysis with the principal emphasis on the *Analytic Hierarchy Process* and *Simos-Roy-Figueira* methods together with some effective improvements of the original approaches to be effectively applied in the construction sectors.

The rationality of this choice concerns the popularity of the approaches and the compatibility with innovative tools. Indeed, the *Analytic Hierarchy Process* method is one of the most used approaches in the construction sector. In addition, both *Analytic Hierarchy Process* and *Simos-Roy-Figueira* have recently been used in combination with *Augmented Reality* and the *User Reporting* to obtain novel tools with great application potential.

1.4 The Multi-criteria Analysis in Building Constructions

The MCDA (or multi-criteria decision making—MCDM) became widely used and effective to assess complex problems and multiple criteria weightings in building construction projects thanks to their adaptability for different purposes and different fields. In the construction sector, MCDA has spread not only for the selection of the best (or most preferred) *solution* among a set of *alternatives*. Indeed, in recent decades, MCDA has been applied in the construction sector also for identifying concise indices or to investigate the human perception.

To this aim, there are three different typologies of goals for the application of MCDA in the area of construction:

Goal A: Definition of *Key Performance Indicators* (KPIs), specific *Condition Ratings* (*Cr*) or risk indices to identify performance, criticality conditions and the risk of the buildings, respectively (Sangiorgio et al. 2020a, b);

Goal B: Identification of the best (or most preferred) solution among a set of *alternatives*;

Goal C: Investigation of stakeholders' *perception.*

All these different purposes to apply MCDA has found application in different phases of the **building process:** *design*, *construction*, *management* and *dismantling* (Schöttle & Arroyo, 2016).

In the *design* phase, MCDA can be used to select the best contractor (Fong & Choi, 2000; Hsieh et al., 2004) or to identify the best material supplier choice (Plebankiewicz and Kubek, 2016; Kahraman et al., 2003). In addition, the most complex and innovative MCDA applications during the construction *design* are applied in the field of intelligent building systems by identifying key selection criteria which include structural, energy, economic, qualitative, aesthetic and multi-risk parameters. These last applications provide adequate solutions by MCDA for the systematic evaluation of many factors including efficiency, user comfort, safety, reliability, functionality and maintainability to characterize the design work and the weighting of soft benefits in comparison to costs (Kwon et al., 2014; Shapira & Goldenberg, 2005) and environmental impact (Chang et al., 2007; Reza et al., 2011).

In the *construction* phase, MCDA can be used to evaluate different executive strategies by simultaneously taking into account time, costs, risks and executive effectiveness. In addition, modern MCDA can be used in combination with *Augmented Reality* (AR) systems in order to support the different stakeholders in the executive phases (Sangiorgio et al. 2021a, b). Some examples concern the use of AR to display directly on site the exact construction procedure.

In the *management* phase, MCDA is employed to support maintenance and intervention by comparing different measures including multi-attribute, multivariate, *qualitative,* and *quantitative* data (Das et al., 2010). In the related literature, such multi-criteria performance assessments are employed to evaluate different aspects of the constructions: safety evaluation and management in construction sites (Dağdeviren & Yüksel, 2008; Li et al., 2013; Teo & Ling, 2006), green building rating (Ali & Nsairat 2009; Chang et al., 2007), energetic rehabilitation (Gigliarelli et al., 2011; Wang et al., 2012). Beyond the analysis of safety and energy aspects, MCDA can also be applied in the field of structural condition assessment (Sangiorgio et al., 2020a, b) including multi-hazard vulnerability analysis by considering seismic (Aghataher et al., 2008; Panahi et al., 2014) and volcanic assessment (Faggiano et al., 2011; Formisano & Mazzolani, 2015). In sum, a fundamental contribution of MCDA, in the *management* phase includes the possibility of considering complex and different phenomena in a comprehensive *multi-risk analysis* to identify the best mitigation strategies for a single building or at the urban and regional scale.

In the *dismantling* phase, similarly to the previous phases, economic, environmental and multi-risk aspects can be considered by including the principles of circular economy and designing out waste and material reuse in MCDA.

On the other hand, MCDA can be difficult to be effectively applied to complex problems in the field of construction because of some limitations of the classical approaches widely recognized in the scientific and technical world (Arroyo et al., 2015; Sangiorgio et al., 2018a, b). Indeed, the main drawbacks are related to:

(1) The need to include *non-expert users* in the analysis to evaluate their needs and perception. Unfortunately, non-expert users are often unable to correctly apply most of the traditional multi-criteria methods.
(2) The *required time* is typically high and the *difficulty* of the analysis conventionally requires trial and error procedures.
(3) The user needs to simultaneously consider *several parameters* (i.e. *criteria*, *sub-criteria* and *alternatives*) and it is demonstrated that the human mind is limited to 7 ± 2 *parameters* (Miller, 1956) to perform effective comparisons.
(4) The need to carry out a large *collection of statistically relevant data* in order to *validate the models*.

Note that in this book, we first explain some classical MCDA methods and then propose and explain in detail some improved MCDA methods supported by specific tools and technologies (such as *User Reporting* and *Augmented Reality*) to overcome conventional MCDA drawbacks.

1.5 The Analytic Hierarchy Process as the Most Used Multi-criteria Approach in Construction

The *Analytic Hierarchy Process* (AHP) provides the objective mathematics to process the inescapably subjective nature of the personal preferences of an individual, the DM, or of a group in making a decision (Saaty & Vargas, 2001). The AHP is based on the decomposition of the problem into *criteria and sub-criteria.* The criteria may or may not be independent. Independence here is defined as what is known in the literature as preferential independence that allows a goal to be represented by a weighted average of the criteria. If they are independent the multidimensional scaling problem is transformed into many one-dimensional scaling problems. Therefore, every *criterion* is analyzed individually in order to identify the related priority vectors, i.e. the weights assigned to each *alternative* or *criterion* (Saaty & Vargas, 2001). The AHP uses the principal eigenvalue method for deriving ratio scale priority vectors from positive reciprocal matrices. In particular, such matrices, named *comparison matrices* or *judgments matrices*, are established through pairs of comparisons (Barzilai et al., 1987; Saaty & Hu, 1998). In addition, a specific *consistency test* is performed on every AHP matrix to verify the coherence of the related pairs comparison.

In the field of construction, *six positive aspects* can be identified in the AHP approach:

(i) The possibility to transform a "multidimensional scaling problem" to many "one-dimensional scaling problems" makes the AHP method particularly *adaptable* in very different contexts. In addition, this characteristic allows involving both *qualitative* and *quantitative* data and *technical* and *non-technical* information in the analysis;
(ii) By using the *Consistency Test*, it is possible to assess the *coherence* of the assigned *judgments* and the relative weights (priorities) obtained;
(iii) The AHP is effective to evaluate the *stakeholder perception* and *users' needs* because the matrices of judgments reflect the subjectivity of the DMs;
(iv) The AHP output provides useful *local* and *global weights* that can be suitable for different purposes including the study of each single problem component (*criterion*, sub *criterion* or *alternative*) or for the overall problem assessment;
(v) The possibility of *aggregating the weights in different ways* (on the basis of several aggregation equations of the related literature) allows performing different types of analyses ranging from the investigation of *stakeholders' perceptions* to complex *multi-risk analyses*;
(vi) The AHP is easy to be implemented in *common spreadsheets (such as Microsoft Excel).*

Thanks to these positive aspects, the AHP has become the most used multi-criteria approach in construction risk management as demonstrated by the literature review of Almeida et al. (2017) (Fig. 1.4).

On the other hand, *six negative aspects* can limit the classical AHP procedure in the construction sector.

Some of these limits are the same as the general drawbacks of the MCDA approach discussed in the previous section. In addition, other negative aspects are specific to AHP applications. All the AHP limits are listed below:

L1. The AHP is difficult to apply for a *large number of criteria* (more than 9) due to the time-consuming required operation and the limits of human mind in analyzing several simultaneous comparisons.

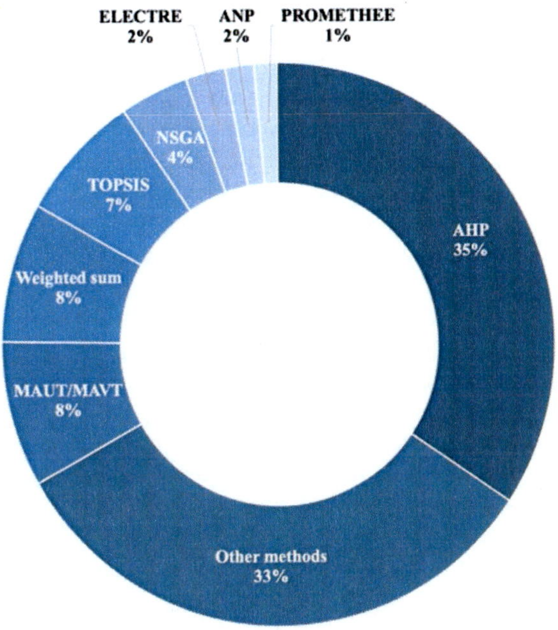

Fig. 1.4 Most used MCA in risk management (Almeida et al., 2017)

L2. The application of the AHP can be *time-consuming* for complex problems involving many *criteria*.

L3 Sometimes it is complex to reach the *Consistency Verification* and the DMs have to perform a trial and error procedure which can compromise the effectiveness of the result.

L4. The *complexity of the decision process* can be a problematic drawback for some inexperienced DM and makes it almost impossible to involve *non-expert users* in the analysis.

L5. A *large data collection* is often necessary to obtain statistically relevant data, the classical AHP is not designed for this possibility.

L6. In some AHP applications, the *subjectivity of judgments* can be a negative aspect that can affect the results. In such cases, an additional *validation procedure* is required.

In the following Sects. (1.6, 1.7 and 1.8) we introduce some of the most recent approaches to support and improve the classical AHP. In particular, a brief description of these new approaches is reported by specifying for each approach what drawbacks of the classic approach are possible to be overcome.

1.6 Simplified Approach for MCDA: The *Simos-Roy-Figueira Method*

In a *Multiple Criteria Decision Analysis* context, knowing the preferences of the DM and determining weights of criteria are very hard problems. Indeed, the large number of *parameters* (i.e. *criteria*, *sub-criteria* and *alternatives*) and the complexity of the decision process can compromise the effectiveness of the analysis. Several methods can be used to give an appropriate value to the weights of *criteria.* Among the existing MCDA method, *Simos-Roy-Figueira's* method is a very simple procedure to accomplish that objective. It consists in using a set of cards to indirectly determine the numerical values of weights (Sangiorgio et al., 2021b). The weighting procedure is obtained by asking the DM to rank the cards (representing the *parameters*) from the least important to the most important (Fig. 1.5). The DM assess the importance of two successive cards (or *parameters*) by introducing white cars. In conclusion, different algorithms can be used to extract the weights from the card ranking. This method was initially developed to be used in the ELECTRE procedure and its potential has recently been demonstrated in combination with different approaches such as the AHP. This procedure was revised by Bernard Roy and José Figueira and it currently is known as the Simos-Roy-Figueira method (SRF) (Figueira & Roy, 2002).

In this book, will be explained how the SRF can be used to improve the AHP and overcome some of the classical drawbacks of the classical approach. In particular, by using this procedure it is possible to consider a large number of *criteria* and perform the analysis by saving time.

Note that this approach overcomes points L1 and L2 of the negative aspects of the AHP discussed in Sect. 1.5.

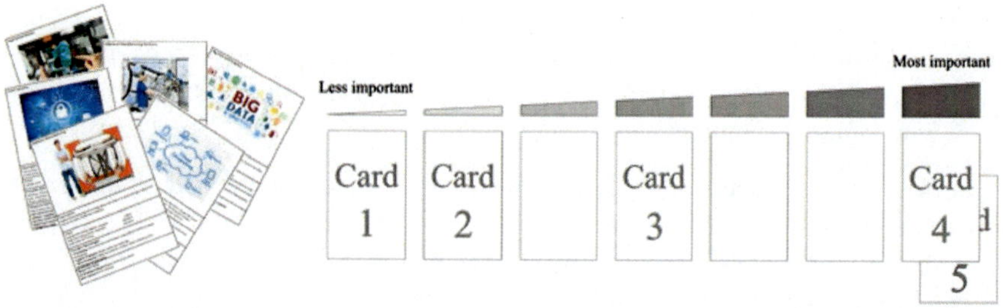

Fig. 1.5 Sets of card and ranking by using the Simos-Roy-Figueira method

1.7 The Augmented Reality to Support MCDA in Constructions

In the field of construction, it is often necessary to involve in the analysis *non-expert users* and to compare a large number of *criteria*. Consequently, many problems are difficult to be solved by the "user" or DM because of the large amount of information to be considered in the weights' evaluation phase. An effective tool that can support multi-criteria methods is *Augmented Reality* (AR). In the literature, there are few attempts to use the AR in support of the decision making process (Pantano et al., 2017). One of these attempts is due to Sangiorgio et al. (2021a, b).who conducted exhaustive studies to synergistically include the *Augmented Reality* technology in multi-criteria decision making settings. This approach to merge MCDA and *Augmented Reality* is not yet widespread even if the potential of AR in the construction sector is amply demonstrated. Indeed, some authors investigated the use of distributed AR in collaborative design applications to support architecture and interior design (Shen et al., 2010). In addition, such tools can be useful in improving construction, management, maintenance and renovation of structures (Yu et al., 2009). Finally, the related literature also demonstrated the potential of some spatial AR helpful at the territorial scale applied to urban designs, geographic information systems and large construction management projects (Feng et al., 2020).

In this book, the state-of-the-art techniques of *Analytic Hierarchy Process* supported by *Augmented Reality* (AR-AHP) is explained in detail to get a powerful and modern multi-criteria decision approach. The AR-AHP exploits the AR virtual environment to visually provide a large amount of information during the decision process. This innovative approach allows to overcome the *complexity of the decision process* and the *Consistency Verification* thanks to the information provided in *Augmented Reality*.

> **Note** that AR-AHP is able to address points L3 and L4 of the negative aspects discussed in the Sect. 1.5.

1.8 The User Reporting Supported MCA in Constructions

To involve a large number of DMs, acquire data form from building users and obtain statistically relevant data useful for the calibration and validation of the results, it is necessary to embed the AHP into a modern large-scale data acquisition process.

In recent decades, the advent of the "Digital Transition" of the building sector brought benefits also in the field of the data acquired from users. Indeed, some authors included advanced techniques of "participatory sensing" to acquire data from the users. The participatory sensing is "*the process whereby individuals and communities use ever-more-capable mobile phones and cloud services to collect and analyze systematic data for use in discovery*". In particular, an effective possibility to bring together the personal experience of the DM and a *huge data collection* acquired from users is offered by the *User Reporting* approach. Indeed, this approach can be supported by technological tools, such as application software (APP) implemented on *Smart Devices* and connected with a Web-Based Platform, to support the data processing and suggest the decision. The *building users* or the DMs can exploit *User Reporting* and the related technological tool to provide useful information that could be used for two different purposes:

(i) To support the weighting process in MCDA; and
(ii) To set a large-scale analysis exploiting *KPIs* or specific *Condition Ratings* resulting from the development of a preliminary MCDA. As already defined (Sect. 1.4), *KPIs* and *Cr* are numerical indicators aimed at quantifying the performance and the criticalities of a construction project, respectively.

In the first case, the *User Reporting* is used to enlarge the set of DMs by asking the users to perform the MCDA through a procedure specifically implemented in the technological tool (e.g. the APP for smart devices).

In the second case, *User Reporting* exploits the results of the MCDA previously developed to apply the outcome on a large-scale (useful for validation, or to investigate several case studies). In this case, considering the example of the APP, the reporting can include photographic inspections taken where the APP users work or live. While acquiring the photographic data, the users are helped by a guided questionnaire in order to complete the data collection.

For the case of the **building process,** *User Reporting* can be integrated in an interactive computer-based system, that helps DMs to utilize data and models to solve unstructured problems (Sangiorgio et al., 2018b). In this field, *User Reporting* can support the maintenance and management of construction at the regional scale (Sangiorgio et al., 2019) in order to perform extensive monitoring of construction safety, even in the absence of scheduled maintenance, but reducing at the same time resources. On the other hand, large-scale maintenance and monitoring procedures present critical points such as incompleteness, non-uniformity of knowledge and necessity of expensive inspection and monitoring. Such issues are very complex for those public or private administrators that possess limited resources and who do not have well-defined methodologies and tools to effectively address surveys and interventions at the regional scale. In the related literature, different approaches are present that use code and protocol guidelines to extensively assess existing buildings' criticalities (Mahli et al., 2012; Brandt & Rasmussen, 2002). Such approaches are based on a semeiotic assessment, by the analysis of damages recognized through visual surveys, which is particularly effective to perform a fast screening of damaged buildings (Yadollahi et al., 2012).

This book explains how to set a user reporting supported decision making in constructions. In addition, some pilot cases are reported to show how this approach has been recently used to investigate several case studies and to validate the MCDA.

User Reporting helps to overcome the drawbacks of *large data collection* and is particularly effective to obtain statistically relevant data needed for a *calibration or validation procedure.*

Note that *User Reporting* is able to address points L5 and L6 of the negative aspects discussed in the Sect. 1.5.

1.9 The Structure and Main Contents of the Book

This book explains four contents from theoretical and practical points of view:

Content (1) AHP (Chaps. 2 and 6);
Content (2) AR-AHP (Chaps. 3 and 5);
Content (3) User Reporting to involve a large number of DMs in an MCDA (Chaps. 4 and 6);
Content (4) User Reporting to apply on a large-scale suitable KPIs, *Cr* or risk index obtained with an MCDA (Chaps. 4 and 7).

In particular, these contents are firstly explained theoretically, secondly through a didactic example and finally by real and complex applications. Beyond the AHP, AR-AHP and *User Reporting*, proposed applications are selected to show the three different uses and goals of MCDA (defined in Sect. 1.4 and reported again as follows):

MCDA Goal A: definition of KPIs, *Cr* or risk index for building construction analysis (Chap. 7);
MDCA Goal B: identification of the best (or most preferred) solution among a set of *alternatives* (Chap. 5);
MCDA Goal C: investigation of stakeholders' *perceptions* (Chap. 6).

Figure 1.6 shows a scheme of the chapters of the book indicating the main content of every chapter and the connections between theoretical matter and the proposed practical applications. A brief description of the chapters of this book is proposed in the following (starting from the current one):

This chapter regards the introduction of the MCDA. This chapter firstly provides an overview of the MCDA, secondly focuses on the most used approach in construction and thirdly introduces the novel techniques to support complex analysis.
Chapter 2 is devoted to the explanation of the AHP. This approach is initially explained from a theoretical point of view including the definitions and classical equations of the method. Afterwards, the AHP is clarified from a practical point of view through a simple and teaching example regarding the risk evaluation in building construction (theoretical content 1).

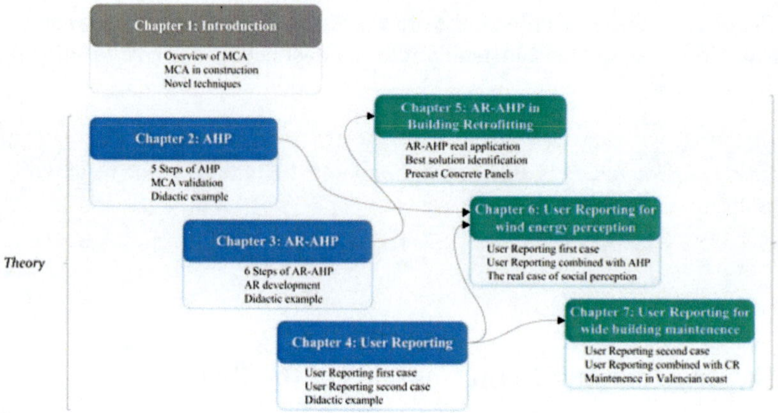

Fig. 1.6 Chapter of the book, principal content and connections between theory and proposed practical application

Chapter 3 presents the AR-AHP, a recent multi-criteria analysis supported by Augmented Reality to perform an MCDA. Also in this case, firstly we explain the theory and the methodological approach. Secondly, the novel AR-based approach is clarified exploiting a practical application with the same didactic example regarding the risk evaluation in building construction (theoretical content 2).

Chapter 4 explains how to set *User Reporting* supported decision making in constructions for two possible different cases. Initially, this chapter focus on the theory regarding the eight steps to set a robust *User Reporting* in construction pointing out the differences with respect to the considered case. Successively, starting from the example of the AHP applied to the risk evaluation in building construction, it is shown how to practically apply the defined eight steps (theoretical content 3 and 6).

The remaining three chapters propose some practical applications of the exceptional academic values in order to demonstrate the potential of the proposed approaches in some real case studies and emphasize the synergy between AHP, *User Reporting* and *Augmented Reality*. In addition, these three applications deal with the three different typologies of goal the multi-criteria analysis respectively: (A) set specific multi-criteria *Condition Ratings*, (B) identify the best (or most preferred) selection among *alternatives* and (C) investigate stakeholders' *perceptions*.

Chapter 5 proposes the application of Augmented Reality based decision making for the identification of the best solution among a set of *alternatives*. In particular, this chapter shows how the AR-AHP supports building retrofitting by identifying the best cladding selection among experimental Precast Concrete Panels-(PCPs). Such application shows the phases of the AR-AHP (in a real building case study) explaining the practical challenges to develop the 3D models for the AR environment, create the decision making in AR and extract weights.

Note that the theoretical concept showed through the application of the AR-AHP and the goal of the MCDA falls into the typology "identification of the best (or most preferred) solution among a set of *alternatives*". The content and MCDA goal showed in this chapter are reported according to the numbering proposed at the beginning of this subsection in the following:

Content 2: AR-AHP;

MCDA Goal B: identification of the best solution among a set of *alternatives.*

Chapter 6 proposes the use of *User Reporting* for the **first case** (involving a large number of DMs in an MCDA). In this case, the reporting is developed in synergy with the AHP to investigate stakeholders' *perceptions.* In more detail, *User Reporting* is used to investigate the perception and the social acceptance of wind energy in Southern Italian cities that are close to wind farms.

Note that this complex application is devoted to showing the synergy and the practical use of the following theoretical content and MCA goal:

Content 1: AHP;

Content 3: *User Reporting* for the **first case** regarding the extension of the set of DMs by obtaining statistically relevant data and create statistical graphs;

MCDA Goal C: MCDA to investigate stakeholders' *perceptions.*

Chapter 7 proposes the use of the *User Reporting* approach for the **second case** (applying on a large-scale a suitable *condition rating* that was preliminary developed). In this case, the AHP is preliminarily used to define specific *Cr* (even already validated) to assess the building state of conservation. In particular, the *User Reporting* is employed to support large-scale building maintenance and diagnostics in the coast of Valencia, Spain.

Note that this complex application is devoted to showing the following theoretical content and MCA goals:

Content 4: *User Reporting* for the **second case** regarding the large-scale analysis to exploit a set of *Cr*;

MCDA Goal A: MCDA to define KPIs, *Cr* or risk indices for building construction analysis.

References

Aghataher, R., Delavar, M. R., Nami, M. H., & Samnay, N. (2008). A fuzzy-AHP decision support system for evaluation of cities vulnerability against earthquakes. *World Applied Sciences Journal, 3*(1), 66–72.

Ali, H. H., & Al Nsairat, S. F. (2009). Developing a green building assessment tool for developing countries–case of Jordan. *Building and Environment, 44*(5), 1053–1064.

Arroyo, P., Tommelein, I. D., & Ballard, G. (2015). Comparing AHP and CBA as decision methods to resolve the choosing problem in detailed design. *Journal of Construction Engineering and Management, 141*(1), 04014063.

Azzabi, L., Azzabi, D., & Kobi, A. (2020). The Multi-criteria approach for decision support. *International Series in Operations Research and Management Science.*

Barzilai, J., Cook, W. D., & Golany, B. (1987). Consistent weights for judgements matrices of the relative importance of alternatives. *Operations Research Letters, 6*(3), 131–134.

Brandt, E., & Rasmussen, M. H. (2002). Assessment of building conditions. *Energy and Buildings, 34*(2), 121–125.

Chang, K. F., Chiang, C. M., & Chou, P. C. (2007). Adapting aspects of GBTool 2005—searching for suitability in Taiwan. *Building and Environment, 42*(1), 310–316.

Dağdeviren, M., & Yüksel, İ. (2008). Developing a fuzzy analytic hierarchy process (AHP) model for behavior-based safety management. *Information Sciences, 178*(6), 1717–1733.

Das, S., Chew, M. Y. L., & Poh, K. L. (2010). Multi-criteria decision analysis in building maintainability using analytical hierarchy process. *Construction Management and Economics, 28*(10), 1043–1056.

de Almeida, A. T., Alencar, M. H., Garcez, T. V., & Ferreira, R. J. P. (2017). A systematic literature review of multicriteria and multi-objective models applied in risk management. *IMA Journal of Management Mathematics, 28*(2), 153–184.

Faggiano, B., Formisano, A., De Gregorio, D., De Lucia, T., & Mazzolani, F. M. (2011). A quick level methodology for the volcanic vulnerability assessment of buildings. In *Applied Mechanics and Materials* (Vol. 82, pp. 639–644). Trans Tech Publications Ltd.

Feng, Z., González, V. A., Amor, R., Spearpoint, M., Thomas, J., Sacks, R., & Cabrera-Guerrero, G. (2020). An immersive virtual reality serious game to enhance earthquake behavioral responses and post-earthquake evacuation preparedness in buildings. *Advanced Engineering Informatics, 45*, 101118.

Figueira, J., Mousseau, V., & Roy, B. (2005). ELECTRE methods. In *Multiple criteria decision analysis: State of the art surveys* (pp. 133–153). Springer.

Figueira, J., & Roy, B. (2002). Determining the weights of criteria in the ELECTRE type methods with a revised Simos' procedure. *European Journal of Operational Research, 139*(2), 317–326.

Fong, P. S. W., & Choi, S. K. Y. (2000). Final contractor selection using the analytical hierarchy process. *Construction Management and Economics, 18*(5), 547–557.

Formisano, A., & Mazzolani, F. M. (2015). On the selection by MCDM methods of the optimal system for seismic retrofitting and vertical addition of existing buildings. *Computers & Structures, 159*, 1–13.

Gigliarelli, E., Cessari, L., & Cerqua, A. (2011, June). Application of the analytic hierarchy process (AHP) for energetic rehabilitation of historical buildings. In *11th International Symposium on the AHP. Wersja elektroniczna.*

Hsieh, T. Y., Lu, S. T., & Tzeng, G. H. (2004). Fuzzy MCDM approach for planning and design tenders selection in public office buildings. *International Journal of Project Management, 22*(7), 573–584.

Kahraman, C., Cebeci, U., & Ulukan, Z. (2003). Multi-criteria supplier selection using fuzzy AHP. *Logistics information management.*

Kwon, S., Lee, G., Ahn, D., & Park, H. S. (2014). A modified-AHP method of productivity analysis for deployment of innovative construction tools on construction site. *Journal of Construction Engineering and Project Management, 4*(1), 45–50.

Li, F., Phoon, K. K., Du, X., & Zhang, M. (2013). Improved AHP method and its application in risk identification. *Journal of Construction Engineering and Management, 139*(3), 312–320.

Mahli, M., Che-Ani, A. I., Abd-Razak, M. Z., Tawil, N. M., & Yahaya, H. (2012). School age and building defects: Analysis using condition survey protocol (CSP) 1 matrix. *International Journal of Civil, Architectural, Structural and Construction Engineering, 6*(7), 56–58.

Miller, G. A. (1956). The magic number seven plus or minus two: Some limits on our capacity for processing information. *Psychological Review, 63*(2), 87–185.

Panahi, M., Rezaie, F., & Meshkani, S. A. (2014). Seismic vulnerability assessment of school buildings in Tehran city based on AHP and GIS. *Natural Hazards and Earth System Sciences, 14*(4), 969.

Pantano, E., Rese, A., & Baier, D. (2017). Enhancing the online decision-making process by using augmented reality: A two country comparison of youth markets. *Journal of Retailing and Consumer Services, 38*, 81–95.

Plebankiewicz, E., & Kubek, D. (2016). Multicriteria selection of the building material supplier using AHP and fuzzy AHP. *Journal of Construction Engineering and Management, 142*(1), 04015057.

Reza, B., Sadiq, R., & Hewage, K. (2011). Sustainability assessment of flooring systems in the city of Tehran: An AHP-based life cycle analysis. *Construction and Building Materials, 25*(4), 2053–2066.

Saaty, T. L., & Vargas, L. G. (2001). How to make a decision. In *Models, methods, concepts & applications of the analytic hierarchy process* (pp. 1–25). Springer.

Saaty, T. L., & Hu, G. (1998). Ranking by eigenvector versus other methods in the analytic hierarchy process. *Applied Mathematics Letters, 11*(4), 121–125.

Sangiorgio, V., Uva, G., Fatiguso, F., & Adam, J. M. (2019). A new index to evaluate exposure and potential damage to RC building structures in coastal areas. *Engineering Failure Analysis, 100*, 439–455.

Sangiorgio, V., Uva, G., & Fatiguso, F. (2018a). Optimized AHP to overcome limits in weight calculation: building performance application. *Journal of Construction Engineering and Management, 144*(2), 04017101.

Sangiorgio, V., Uva, G., & Fatiguso, F. (2018b). User reporting–based semeiotic assessment of existing building stock at the regional scale. *Journal of Performance of Constructed Facilities, 32*(6), 04018079.

Sangiorgio, V., Uva, G., & Aiello, M. A. (2020a). A multi-criteria-based procedure for the robust definition of algorithms aimed at fast seismic risk assessment of existing RC buildings. In *Structures* (Vol. 24, pp. 766–782). Elsevier.

Sangiorgio, V., Silvia, M., & Fatiguso, F. (2020b). Augmented reality to support multi-criteria decision making in building retrofitting. In 2020 IEEE International Conference on Systems, Man, and Cybernetics (SMC) (pp. 760–765). IEEE.

Sangiorgio, V., Martiradonna, S., Fatiguso, F., & Lombillo, I. (2021a). Augmented reality based-decision making (AR-DM) to support multi-criteria analysis in constructions. *Automation in Construction, 124*, 103567.

Sangiorgio, V., Di Pierro, B., Roccotelli, M., Silvestri, B. (2021b) "Card game analysis for fast multi-criteria decision making" *RAIRO - Operations Research.*

Schöttle, A., & Arroyo, P. (2016). The impact of the decision-making method in the tendering procedure to select the project team. In *Proceeding of 24th Annual International Group for Lean Construction, IGLC East Lansing, MI* (pp. 23–32).

Shapira, A., & Goldenberg, M. (2005). AHP-based equipment selection model for construction projects. *Journal of Construction Engineering and Management, 131*(12), 1263–1273.

Shen, Y., Ong, S. K., & Nee, A. Y. (2010). Augmented reality for collaborative product design and development. *Design Studies, 31*(2), 118–145.

Teo, E. A. L., & Ling, F. Y. Y. (2006). Developing a model to measure the effectiveness of safety management systems of construction sites. *Building and Environment, 41*(11), 1584–1592.

Wang, E., Shen, Z., Neal, J., Shi, J., Berryman, C., & Schwer, A. (2012). An AHP-weighted aggregated data quality indicator (AWADQI) approach for estimating embodied energy of building materials. *The International Journal of Life Cycle Assessment, 17*(6), 764–773.

Yadollahi, M., Adnan, A., & Zin, R. M. (2012). Seismic vulnerability functional method for rapid visual screening of existing buildings. *Archives of Civil Engineering, 58*(3), 363–377.

Yu, D., Jin, J. S., Luo, S., Lai, W., & Huang, Q. (2009). A useful visualization technique: a literature review for augmented reality and its application, limitation & future direction. In *Visual information communication* (pp. 311–337). Springer.

Chapter 2
The Analytic Hierarchy Process in the Building Sector

Abstract The *Analytic Hierarchy Process* (AHP) is one of the most used multi-criteria approaches in the **architectural engineering** and **building production** tanks to its possibility to transform a "multidimensional scaling problem" to many "one-dimensional scaling problem". This characteristic makes the AHP method particularly *adaptable* in very different contexts and allows to involve both *qualitative* or *quantitative* data and *technical* or *non-technical* information in the analysis. In this chapter, the fundamentals of the *Analytic Hierarchy Process* and practical applications in the building sector are explained. In particular, the five steps necessary to develop a multi-criteria decision analysis (MCDA) with the AHP are described in detail from both a theoretical and practical points of view: (i) *Structuring of the problem with a hierarchy*, (ii) *Local weights evaluation*, (iii) *Local consistency test*, (iv) *Global weights evaluation and*, (v) *Sensitivity analysis*. Indeed, **firstly** theoretical aspect of the five steps is explained and discussed in specific subsections of this chapter. **Secondly**, additional subsections clarify the same theoretical concepts from a practical point of view by exploiting a basic example concerning a simplified risk evaluation in the field of building construction. To this aim, a guided procedure explains how to implement the AHP in a spreadsheet.

Keywords Analytic Hierarchy Process · Architectural engineering · Building production · Theoretical concepts · Practical application · Risk evaluation · Spreadsheets

In this chapter, we explain the fundamentals of the *Analytic Hierarchy Process* and practical applications in the building sector. The theoretical introduction of the method is presented following the footstep of the original Saaty's (Saaty & Vargas, 2012a) discussion of AHP or Brunnelli's (2014). In particular, the five steps necessary to develop an MCDA model with the AHP are shown in the flowchart of Fig. 2.1 and listed as follows: (i) *Structuring of the problem with a hierarchy*, (ii) *Local weights evaluation*, (iii) *Local consistency test*, (iv) *Global weights evaluation and*, (v) *Sensitivity analysis*. Every one of these steps is explained and discussed in a specific subsection of this chapter. Moreover, after the theoretical explanation of the AHP, useful simplified equations to implement the approach in spreadsheets are explained.

V. Sangiorgio et al., *New Approaches for Multi-Criteria Analysis in Building Constructions*, https://doi.org/10.1007/978-3-030-83875-1_2

Fig. 2.1 Flowchart of the classical AHP

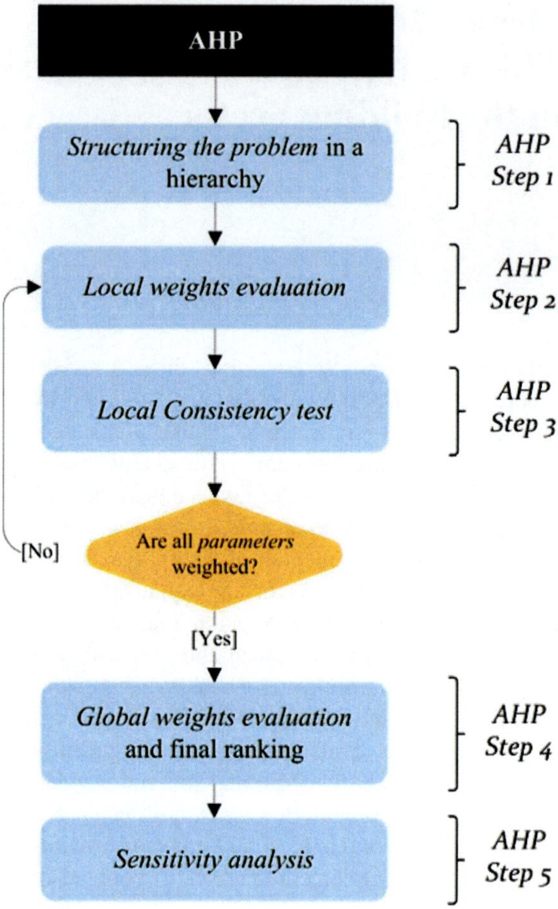

In addition, a further subsection explains the same theoretical concepts from a practical point of view by exploiting a basic example. The example concerns a simplified risk evaluation in the field of building construction risk management.

2.1 Structuring of the Problem with a Hierarchy

Starting with a decision problem, the first step consists of structuring the problem according to a hierarchical scheme, to provide a detailed, simple, systematic and structured decomposition of the general problem into its basic components. To this aim, the goal of the AHP is identified and the related *criteria*, *sub-criteria* and *alternatives* to reach the goal are determined (Fig. 2.2).

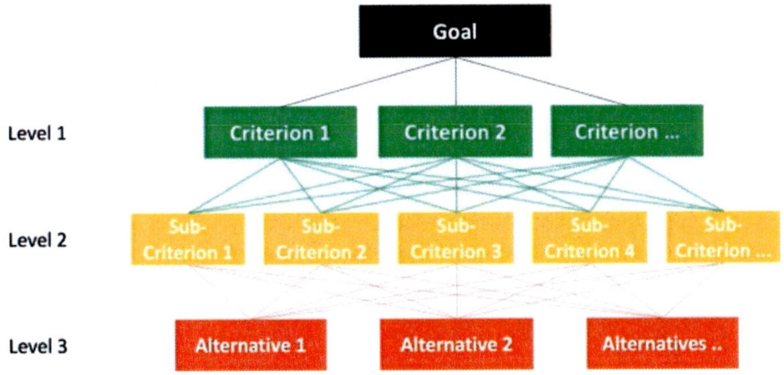

Fig. 2.2 Generic flowchart of *the structuring of the problem in a hierarchy*

Note that the *precision* about the aim of the AHP helps to correctly define the problem structure and keeps the analysis on track. Indeed, first of all, it is fundamental to correctly answer the following question: **What is the purpose of your AHP?** Get this wrong and you can provide a helpful analysis for the wrong problem. A correct and focused statement of initial aims is crucial to effectively formulate the successive stages.

Once the goal is correctly set, the following five-points guideline to correctly identify *criteria*, *sub-criteria* and *alternatives* can be used:

(i) *Completeness*—Have all important *criteria* been included? Answering this question requires a deep study of the problem since in some cases it is not necessarily obvious from the beginning what the important *criteria* are.

(ii) *Redundancy*—Are there *criteria* which are unnecessary? It is recommended to remove all unnecessary *criteria.*

(iii) *Operationality*—It is important that any *sub-criterion* or alternative can be judged with respect to the relative *criterion.*

(iv) *Optionally*—The involved *parameters* (*criteria*, *sub-criteria* or *alternatives*) may be quantitative, with respect to some commonly shared and understood scale of measurement (like weight or distance). On the other hand, the parameters can be also qualitative, reflecting the subjective assessment of an expert.

(v) *Size*—An excessive number of *criteria*, *sub-criteria* or *alternatives* leads to a problem that is too complex to be able to achieve a coherent result. In this case, the analysis can be very complex and can require improved versions of the classical AHP. Mathematical and psychology studies determined that the human mind is limited to 7 ± 2 alternatives in simultaneous comparison of the classical AHP (Miller, 1994). To this aim, it is recommended to not overcome this number of *criteria*, *sub-criteria* (related with the same *criterion*) or *alternatives*. If a large number of criteria is required, they can be grouped into clusters using some common dimensions. If it is inevitable that the number of criteria is large, a linking procedure can be used to prioritize the criteria without having to compare all of them. The criteria are grouped into 7 ± 2 groups where any two groups must have a criterion in common that serves as the link between the priorities. Thus, to join to sets of priorities the priority of the linking element must be the same in both groups. The priorities are then renormalized for them to add up to unity.

Note that in both the scientific or technical applications of the AHP it is important that the defined *parameters* are effectively justified and/or corroborated by an exhaustive bibliographic investigation in the related literature.

2.2 The Local Weights Evaluation

The second step, i.e. the *local weights evaluation,* is the core of the method and provides the weights that are necessary for generating the ranking. More precisely, it is possible to individually analyze each aspect of the decision problem. Considering n ordered *parameters* for comparison (i.e. *criteria, sub-criteria* or *alternatives* in relation with *criteria, sub-criteria* or *alternatives* or *sub-criteria*), an $n \times n$ *judgments matrix* A is defined (Fig. 2.3), where each upper diagonal element $a_{ij} > 0$ is generated by comparing the $i^{\text{-th}}$ element with the $j^{\text{-th}}$ one through the fundamental scale of absolute numbers. The number of required judgments (rg) to achieve every matrix, depends on the number of the involved *parameters* n and consequently the dimension of the matrix according with the following equation:

Fig. 2.3 Generic *Judgment Matrix A*

A	1	2	…	n
1	1	$a_{1,2}$	…	$a_{1,n}$
2	$1/a_{1,2}$	1	…	$a_{2,n}$
…	…	…	1	…
n	$1/a_{1,n}$	$1/a_{2,n}$	…	1

$$\mathrm{rg}(n) = \frac{n(n-1)}{2} \tag{2.1}$$

This semantic scale is composed of verbal scales that are associated with numerical values (1, 3, 5, 7, 9) and compromises (2, 4, 6, 8) between such values (Table 2.1) (Saaty, 2008).

The lower triangular part of the matrix A is completed with the reciprocal values of the upper triangular part, by setting elements in the transposed location of the matrix to the reciprocal value: $a_{ji} = \frac{1}{a_{ij}}$ or $a_{ji} \cdot a_{ij} = 1$. Moreover, if $a_{ij} \cdot a_{jk} = a_{ik}$, for all i, j and k, then the matrix A is said to be *consistent* and its principal eigenvalue is $\lambda_{\max} = n$.

In the standard AHP, the weights are obtained by solving the following eigenvector problem:

Table 2.1 Fundamental scale of absolute numbers of Saaty

Number value a_{ij}	Verbal scale	Explanation
$a_{ij} = 1$	Equal importance	Two activities contribute equally
$a_{ij} = 3$	Moderate importance of one over another	Experience and judgment slightly favour one activity over another
$a_{ij} = 5$	Strong importance	Experience and judgment strongly favour one activity over another
$a_{ij} = 7$	Very strong importance	An activity is favoured very strongly over another
$a_{ij} = 9$	Extreme importance	An activity is favoured by at least an order of magnitude
1.5–4–6–8	Intermediate value	Used a compromise between two judgments
The reciprocal number express an opposite judgment		*Experience and judgment dominates one activity by another*
$a_{ij} = 1/3$	Moderate less importance of one over another	Experience and judgment slightly dominates one activity by another
$a_{ij} = 1/5$	Strong less importance	Experience and judgment strongly dominates one activity by another
$a_{ij} = 1/7$	Very strong less importance	An activity is dominated very strongly by another
$a_{ij} = 1/9$	Extreme less importance	An activity is dominated by at least an order of magnitude

$$Aw = \lambda_{\max} w. \tag{2.2}$$

where w is the principal right eigenvector associated with the principal eigenvalue $\lambda_{\max}$. If slight inconsistencies are introduced, then it $\lambda_{\max} > n$ (Saaty, 2003).

Note that the AHP can include also quantitative parameters in the analysis. If the DM needs to evaluate a set of n ordered *criteria* measured in the same measurement scale, i.e. numerical values with the same unit of measurement are available, the elements a_{ij} can be directly obtained by forming the ratio between numerical values of the parameters.

2.2.1 *The Simplified Approach for Weights Evaluation*

Operatively, approximate formulation methods are used to calculate the weights (priorities) from the judgment matrix. Saaty and Vargas (1984) showed that the limit of the average of the normalized columns of a positive reciprocal matrix converges to the principal right eigenvector of the matrix. In this chapter, we provide the first order approximation given by Eq. (2.4). These equations are suggested because they can be easily and effectively implemented in spreadsheets for the weights' evaluation of the AHP.

First, the elements of matrix A can be normalized as follows:

$$x_{ij} = \frac{a_{ij}}{\sum_i a_{ij}} \quad \text{for} \quad j = 1, \ldots, n. \tag{2.3}$$

Second, the weight w_i are calculated as the average of the elements of the rows of the normalized matrix:

$$w_i = \sum_{j=1}^{n} \frac{x_{ij}}{n} \quad \text{for} \quad i = 1, \ldots, n. \tag{2.4}$$

2.3 The Local Consistency Test

In the AHP, a suitable consistency test can be performed on every matrix to understand if the judgments have been assigned with consistency. To this aim, Saaty defined the consistency index CI to check the consistency of the judgments assigned. Such an index increases proportionally with the inconsistency of the matrix:

Table 2.2 Noble's random consistency index (Noble & Sanchez, 1993)

n	1	2	3	4	5	6	7	8	9	10	11	12	13	14	15
RI(n)	0	0	0.49	0.82	1.03	1.16	1.25	1.31	1.36	1.39	1.42	1.44	1.46	1.48	1.49

$$\mathrm{CI} = \frac{\lambda_{\max} - n}{n - 1}. \tag{2.5}$$

Consistency is a necessary condition when making decisions, but it is no sufficient for validity. To implement the idea of consistency, Saaty suggested to use a ratio of the consistency index (CI) and the same index but for randomly generated matrices denoted as RI(n) denoted as the Consistency Ratio:

$$\mathrm{CR} = \frac{\mathrm{CI}}{\mathrm{RI}(n)} \tag{2.6}$$

Saaty selected RI(n) to be the average of the CI's for a large number of matrices whose entries were randomly generated from his 1–9 absolute scale. The idea of the consistency ratio index (CR) is to test if the judgments given in a matrix are given at random or thinking about the problem at hand. It was never intended to be a statistical test. For more details about the difference between a statistical test of consistency and Saaty's consistency ratio test see Vargas (2008).

Among the different values of RI proposed in the literature, those in Noble and Sanchez (1993) were used, as reported in Table 2.2. On the basis of several empirical studies, Saaty concluded that the value of Consistency Ratio CR < 0.10 is acceptable (Saaty, 2008).

2.3.1 *The Simplified Approach for the Consistency Test*

It is possible to approximate the principal eigenvalue by using the first order approximation of the principal right eigenvector (Ishizaka & Lusti, 2006). Initially, it is necessary to evaluate the normalized consistency vector **cv*i*** with elements cv$i$ obtained by the following equation (particularly useful to be implemented in a spreadsheet):

$$\mathrm{cv}_i = \sum_{j=1}^{n} \frac{a_{ij} \times w_i}{w_i} \quad \text{for} \quad i = 1, \ldots, n. \tag{2.7}$$

Afterwards, $\lambda_{\max}$ can be evaluated as the average of the elements of vector **cv*i*** according to the following equation:

$$\lambda_{\max} = \frac{1}{n} \sum_{i=1}^{n} \sum_{j=1}^{n} \frac{a_{ij} \times w_i}{w_i} = \frac{1}{n} \sum_{i=1}^{n} cv_i. \tag{2.8}$$

2.4 Global Weights Evaluation

The *global weights evaluation*, the synthesis of priorities (or weights aggregation), is performed to determine the rankings and the global weights for each alternative. There are two situations to consider: (1) hierarchical composition—when the criteria weights are independent of the alternatives and (2) network composition—the criteria weights depend on the alternatives. To this aim, the local weights of every level of the *the structuring of the problem in a hierarchy* (see Fig. 2.2) are aggregated together in order to obtain the global weights at each level. In the existing literature, there are many ways of aggregating the local weights and the use of a specific equation depends on the problem investigated (Sangiorgio et al., 2018, 2019, 2020a, 2020b). The simplest and widespread weights aggregation existing in the literature is hierarchical composition or "weighted sum". In such aggregation, the global weights are obtained by multiplying the weights of macro-criteria with weights of criteria and sub-criteria and finally totalling the results for each alternative according to the connections created in the hierarchical *structure of the problem* (Sangiorgio et al., 2021; Harker & Vargas, 1987). In addition, the classical AHP allows the weights of multiple decision makers to be combined by synthesizing the individual judgments of several experts or users. This approach, as shown by Saaty and Vargas (2012b), bypasses the problem created by Arrow's Impossibility Theorem.

Note that in risk analysis we use a factorial equation to aggregating the parameters. It is based on the product of the weights related to the *criteria* involved.

2.4.1 *Global Weights Evaluation in a Generic Example*

To provide an example of the *global weights evaluation* and final aggregation, let us consider the generic structure of the problem in a hierarchy shown in Fig. 2.4. In this problem, three criteria, three sub-criteria and two alternatives are defined. In this example, the *structure of the problem* is defined with the following connections.

Sub-criterion 1 is connected with *criterion 1* and *criterion 2*;
Sub-criterion 2 is connected with *criterion 1* and *criterion 2*;
Sub-criterion 3 is connected with *criterion 2* and *criterion 3*;
Alternative A is connected with *Sub-criterion 1* and *Sub-criterion 2*;
Alternative B is connected with *Sub-criterion 2* and *Sub-criterion 3*.

In addition, the *f* local weights can be calculated using the equations given in Sect. 2.2.

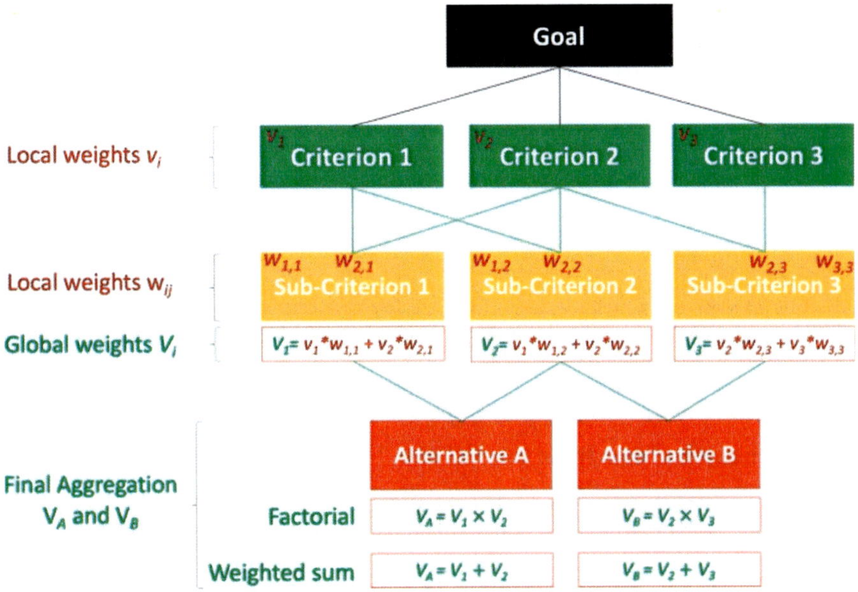

Fig. 2.4 Generic flowchart of *the structuring of the problem in a hierarchy*

Let v_1, v_2, v_3 be the weights associated with *criterion 1*, *criterion 2* and *criterion 3*, respectively;

Let $w_{1,1}$ and $w_{1,2}$ be the weights associated with *Sub criterion 1* and *Sub-criterion 2*, respectively, in relation to *criterion 1*;

Let $w_{2,1}$, $w_{2,2}$ and $w_{2,3}$ be the weights associated with *Sub-criterion 1, Sub-criterion 2* and *Sub-criterion 3*, respectively, in relation to *criterion 2*; and

Let $w_{3,3}$ be the weight associated with *Sub-criterion 3*.

Once the local weights are evaluated, the global weights V_1, V_2 and V_3 of the *sub-criteria* can be calculated by multiplying the weights of a *criterion* with the weights of the *sub-criteria* totalling the results according to the connections created in the *structure of the problem* as follows:

$$V_1 = v_1 * w_{1,1} + v_2 * w_{2,1}$$

$$V_2 = v_1 * w_{1,2} + v_2 * w_{2,2}$$

$$V_3 = v_2 * w_{2,3} + v_3 * w_{3,3}.$$

In conclusion, the weighs can be aggregated by using the factorial formula or by using the weighted sum to obtain the global weights V_A and V_B of the *alternatives* (useful to obtain the final ranking).

Factorial formula:

$$V_A = V_1 V_2 = (v_1 * w_{1,1} + v_2 * w_{2,1}) \times (v_1 * w_{1,2} + v_2 * w_{2,2})$$

$$V_B = V_2 V_3 = (v_1 * w_{1,2} + v_2 * w_{2,2}) \times (v_2 * w_{2,3} + v_3 * w_{3,3})$$

Weighted sum:

$$V_A = V_1 + V_2 = v_1 * w_{1,1} + v_2 * w_{2,1} + v_1 * w_{1,2} + v_2 * w_{2,2}$$

$$V_B = V_2 + V_3 = v_1 * w_{1,2} + v_2 * w_{2,2} + v_2 * w_{2,3} + v_3 * w_{3,3}$$

2.5 Sensitivity Analysis

The final step of the procedure is the *sensitivity analysis* that shows how the final results would change if the weights of the *parameters* were different. It also helps to understand how robust (or stable) the final weights are and which *parameters* have greater influence on the results. The *sensitivity analysis* is performed by changing the weights of the *criteria, sub-criteria* or *alternatives* and verifying how they modify the overall priorities of the alternatives. It is worth noting that this additional analysis is not always necessary and depends on the problem under consideration.

Also in this case, as for the *global weights evaluation*, there are different typologies of *sensitivity analysis* in the related literature. There is no general rule to select a specific procedure, but it is important to understand which part of the decision process should be investigated more in dept and why.

In particular, the following part of the decision problem could be investigated (but not only):

(i) At the level of *criteria*, it could be interesting to understand if the final ranking change by modifying some criteria weights or by including other/or different *criteria*;
(ii) At the level of the *sub-criteria* it is possible to understand if the final ranking change by modifying some *sub-criteria* weights or by including other/or different *sub-criteria*;
(iii) At the level of the *alternatives* it is possible to understand if the final ranking change by modifying some *alternatives* weights with respect to the criteria and sub-criteria or by including other/or different *alternatives*;
(iv) At the level of the *decision makers* if the results are obtained by combining many DMs. In this case, it could be interesting to understand if the final ranking suffers important changes varying the DMs involved.

Using the *sensitivity analysis* to examine how the ranking of options might change under different scoring or weighting systems can confirm that the final result is robust and does not change even if small input data change. Alternatively, the analysis can show that two or three options always come out best, but their order may shift. In this last case, if the differences between these best options under different weighting systems are small, then accepting a second-best option can be shown to be associated with little loss of overall benefit.

Note that the sensitivity analysis is neither calibration nor validation but can be used to verify the robustness of the final result in relation to small changes of the application of the procedure.

2.6 Calibration and Validation of the Output of the AHP

Calibration and Validation are fundamental in academic and technical studies to give scientific soundness to an investigation. Unfortunately, often when a problem is studied with an MCDA, these steps may be neglected and completely absent. This section intends to provide an overview of Calibration and Validation procedures to provide a useful guide in selecting the best strategy in relation to the problem considered.

In this subsection, first, we define calibration and validation in the case of an MCDA and then we provide an overview of possible validation and calibration approaches that could be applied to the MCDA in question.

Calibration is the operation that establishes a relation between the results obtained from an MCDA under test (to be calibrated) with those of a different approach of known accuracy (or based on experimental and real data). The purpose of the calibration is to reduce the differences between the result of the MCDA and the data obtained with the different approaches of known accuracy. For instance, in an MCDA the calibration can be performed on weights that are not robust or are difficult to be obtained with expert judgments. Such weights can be left as variables and then mathematically calibrated (Sangiorgio et al., 2020a).

Validation is the operation that involves laboratory tests, case study application and/or comparisons to verify that the results delivered by an MCDA under test (to be validated) are working properly (i.e. effectively reflects what happens in reality).

Note that an example of *calibration* and validation of an MCDA can be found in the open access research of Sangiorgio et al. (2020a). In this work, an MCDA is used to obtain an index to forecast the natural phenomenon of the Urban Heat Island (UHI) and the calibration is performed to reduce the differences between the result of the MCDA and the real UHI measured in 32 urban districts. The calibration in this case is focused on the weights of the criteria that cannot be easily and/or effectively obtained from expert judgments. In addition, in the same work, the validation is achieved by verifying that the obtained index reflects what happens in reality (evaluating absolute and relative error).

In what follows we provide an overview of some of the possible approaches to *calibrate* and *validate* an MCDA.

The first important consideration concerns the goal typology of the MCA (remember the definition in Sect. 1.4) and the typology of the output of the procedure. The goal typology of the MCDA and the subsequent output could be as follows:

MCDA Goal A: The MCDA devoted to quantifying an *index* can yield KPIs, a *Cr* or Risk indices.

- The KPIs or *Cr* can be dimensionless quantities or can have a physical dimension;
- Risk indices typically are dimensionless.

MCDA Goal B: The identification of the best solution among a set of *alternatives* provides a ranking expressed with *Natural numbers* (1st, 2nd, 3rd and so on).

MCDA Goal C: The investigation of the stakeholders' *perceptions* can achieve a *ranking* or directly the *weights of the parameters.*

- The *ranking* of the parameters is expressed with *Natural numbers.*
- The *weights of the parameter* are typically *normalized values* that can be expressed in percentages (%).

According to the previous association between goals and outputs, four different typologies of outputs can be achieved with a generic MCDA:

(1) *Physical dimensions*;
(2) *Dimensionless quantities*;
(3) *Natural numbers; and*
(4) *Normalized values (*e.g.*%).*

In conclusion, a specific *calibration* or *validation* procedure can be applied to every output typology as follows:

(1) An MCDA providing a *Physical dimension* can be calibrated with *regression analysis,* a *mathematical optimization* model exploiting experimental, statistical and real data. In addition, the validation can be performed by using measures such as *absolute error* and *relative error* or by performing a *correlation* analysis (Sangiorgio et al., 2020a). On the other hand, other approaches such as comparisons and empirical validation can be applied, but are not recommended for these types of MCDA and output.
(2) An MCDA specifying a *Dimensionless quantity* can be *calibrated* in a way similar to the case of physical dimensions. On the other hand, in the *calibration* and *validation processes* it could be difficult to find real data that can be used to verify that the approach is working properly. To this aim, a common validation can be performed by *comparing* the MCDA results with more complex or

similar approaches (Sangiorgio & Parisi, 2020; Sangiorgio et al., 2019). In absence of the possibility of using mathematical and statistical calibration, or comparison with more effective techniques, it is possible to perform the validation with other less rigorous approaches. For example, by considering a *significant statistical number of* DMs to reach a shared and uniform decision based on experts' judgments. In this last approach, we can only obtain an indicative suggestion of the quality of the results.

(3) An MCDA which gets a result in the form of *natural numbers* (representing the ranking of parameter or alternatives) can only be *calibrated* by involving a large number of experts (DMs) for a shared final choice. The *validation* can exploit *empirical validation* based on a *large number of applications* and by performing a *statistical analysis of the ranking reversal* based on a large set of involved DMs.

(4) An MCDA yielding *normalized values* (typically used to investigate *perception*) can be *calibrated* by involving a *statistical significant number of* DMs and *validated* empirically considering a *large number of applications* (Caporale et al., 2020).

Figure 2.5 shows a diagram associating the *Goal of an MCDA*, the relative *output* typology and the related recommended approach to *calibrate* and *validate* the results.

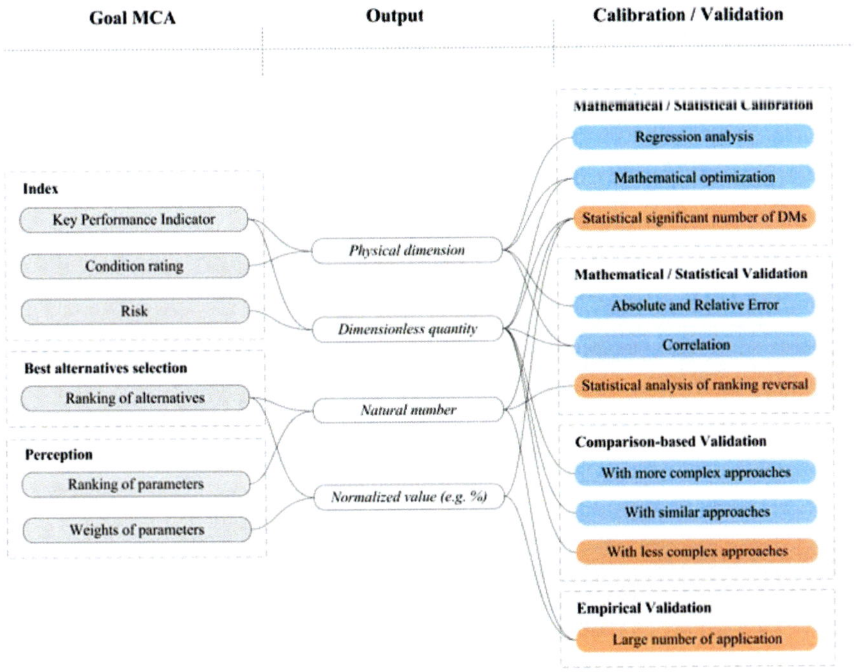

Fig. 2.5 Relation among the goal of the MCDM obtained output and available approaches for calibration and validation

2.7 Example: A Simplified Risk Evaluation in the Field of Building Construction Management

In this subsection, a simple explanation of the proposed approach from a practical point of view is given to better clarify the use of the AHP by exploiting simple spreadsheets.

In order to present a multi-criteria problem in the field of the construction *management* phase, this chapter presents the case of the heavy expulsion of concrete cover and rebars oxidation in the residential building sector. In particular, a simplified MCDA regarding the Risk for concrete blocks falling down is specifically developed for this educational purpose. Figure 2.6 shows two buildings affected by heavy expulsion of concrete cover. In addition, on the right side of Fig. 2.6 a stylized image shows the type of risk to be analyzed: the risk of injury to the users due to the detachment of the building concrete cover.

Note hat the proposed example contemplates simultaneously two MCDA typologies of goal (*A* and *B*) defined in the Sect. 1.4. Indeed, in this simplified analysis, a risk index (*A*—definition of KPIs, *CR* or risk index for building construction analysis) is calculated to obtain a ranking of three alternatives (*B*—identification of the best solution among a set of *alternatives*).

2.7.1 Example: The Structuring of the Problem

In particular, the multi-criteria decision problem concerns the identification of which of the three situations (a, b, c) shown in Fig. 2.7 is the riskiest.

Fig. 2.6 Risk evaluation in the field of building construction management

The three ***alternatives*** are: (a) *two men* with *sombrero hat* and block falling from 3 m, (b) *one man* with *helmet* and block falling from 6 m and (c) *three children* without head cover (*nothing*) and block falling from 1 m.

As mentioned before, the question is: *which is the riskiest situation among a, b and c?* The answer to this question is not necessarily obvious and consequently, a Multi-Criteria Analysis such as AHP is fundamental to analyze the problem and identify the riskiest situation.

The ***objective*** can be summarized in the "***Risk of falling objects***". Moreover, in accordance with the existing risk assessment procedure (IEC, 2008; Towhata, 2008; Sangiorgio & Parisi, 2020) three well-known risk criteria are used and customized for the phenomenon considered: *Hazard*, *Vulnerability* and *Exposure*.

The ***criteria*** are defined as follows:

The *Hazard* represent the *height of the fall* of the concrete cover that can be 3 m, 6 m or 1 m.

The *Vulnerability* depends on the *head cover* (*Sombrero hat, helmet* or *nothing*).

The *Exposure* is related to the type of users (*two men, one man, three children*).

At this point in accordance with the AHP first step of the *structuring of the problem in a hierarchy* it is possible to draw the flowchart of the ***Risk of falling objects*** (Fig. 2.8).

Note that in the flowchart, every alternative is connected only with the related *sub-criteria* (e.g. A is connected with *Sombrero hat*, 3 m and *two man*).

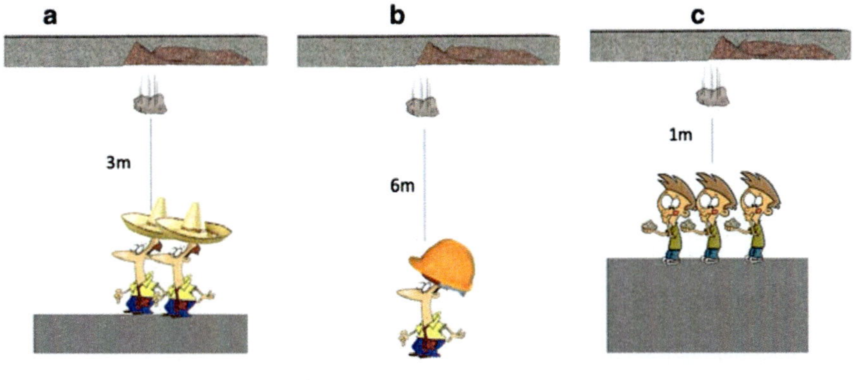

Fig. 2.7 Three situations of risk: **a** two men with Sombrero hat and block falling from 3 m, **b** one man with helmet and block falling from 6 m, **c** three children without head cover and block falling from 1 m

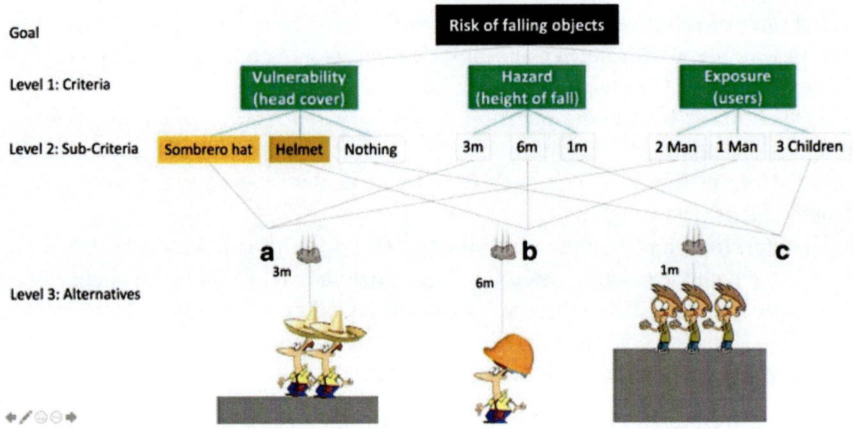

Fig. 2.8 Flowchart: structuring of the "Risk of falling objects" problem in a hierarchy

2.7.2 Example: The Local Weights Evaluation

Starting from the first *criterion* of *Vulnerability* (Fig. 2.9) it is possible to obtain the local weights of the three *sub-criteria* of *sombrero hat, helmet or nothing*. To this aim the matrix A_1 can be obtained by performing pairwise comparisons.

In every cell of the *Judgment Matrix* (Fig. 2.10), the value a_{ij} can be assigned by the DMs by using the numerical scale shown in Table 2.1 reflecting their preference (also called judgment). In particular, in this case, the DM need to evaluate qualitatively (based on his experience) how much a *sub-criterion* is more "vulnerable" than *another*. In order to make the understanding of the approach easier, Table 2.1 has been customized to the example of the *Vulnerability criterion* and showed as follows (Table 2.3).

For the first *criterion*, the DM express the following evaluation (considering the upper triangular part of the matrix):

Sombrero is much more "vulnerable" than *Helmet* $a_{1,2} = 5$ $a_{2,1} = 1/5$
Sombrero is moderately less "vulnerable" than *nothing* $a_{1,3} = 1/3$ $a_{3,1} = 3$
Helmet is extremely less "vulnerable" than *nothing* $a_{2,3} = 1/9$ $a_{3,2} = 9$

Figure 2.11 shows the *Judgment Matrix* A_1 related to the *criterion Vulnerability* (head cover) with the values of the pairwise comparisons.

Once the judgments are obtained, it is possible to implement the simplified approach in spreadsheets such as Excel.

First, the elements of the column are summed and the normalized matrix is obtained by using Eq. (2.3).

Second, considering the normalized matrix, the weights are obtained by averaging the elements of the rows according to Eq. (2.4) (Fig. 2.12).

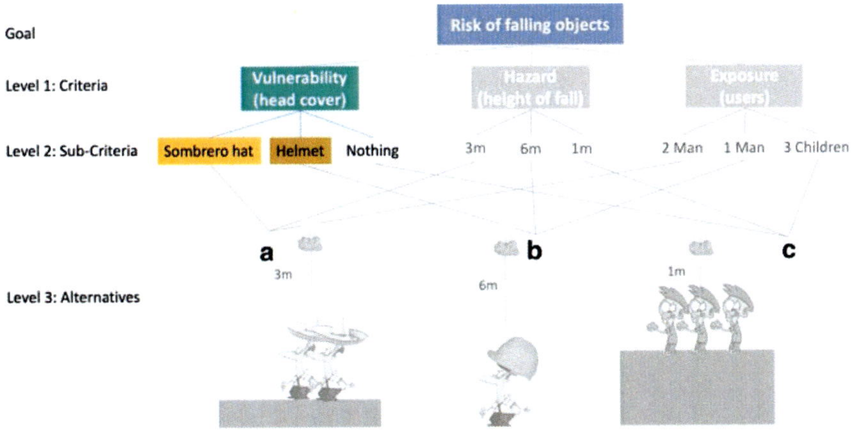

Fig. 2.9 Structuring of the problem: first *criterion* and relative *alternatives*

Matrix A_1: (head cover)	Sombrero	Helmet	Nothing
Sombrero	1	$a_{1,2}$	$a_{1,3}$
Helmet	$a_{2,1}$	1	$a_{2,3}$
Nothing	$a_{3,1}$	$a_{3,2}$	1

Fig. 2.10 *Judgment Matrix* A_1 related to the *criterion Vulnerability* (head cover)

Table 2.3 Fundamental scale of absolute numbers of Saaty adapted for the *Vulnerability criterion*

Number value a_{ij}	Adapted verbal scale
$a_{ij} = 1$	Equally "vulnerable"
$a_{ij} = 3$	Moderately more "vulnerable"
$a_{ij} = 5$	Strongly more "vulnerable"
$a_{ij} = 7$	Very strongly more "vulnerable"
$a_{ij} = 9$	Extremely more "vulnerable"
Number value a_{ij}	Reciprocal number express an opposite judgment
$a_{ij} = 1/3$	Moderately less "vulnerable"
$a_{ij} = 1/5$	Strongly less "vulnerable"
$a_{ij} = 1/7$	Very strongly less "vulnerable"
$a_{ij} = 1/9$	Extremely less "vulnerable"

Matrix A_1: (head cover)	Sombrero	Helmet	Nothing
Sombrero	1	5	1/3
Helmet	=1/5	1	1/9
Nothing	3	9	1

Fig. 2.11 *Judgment Matrix A_1* related to the *criterion Vulnerability* (head cover) and judgment expressed

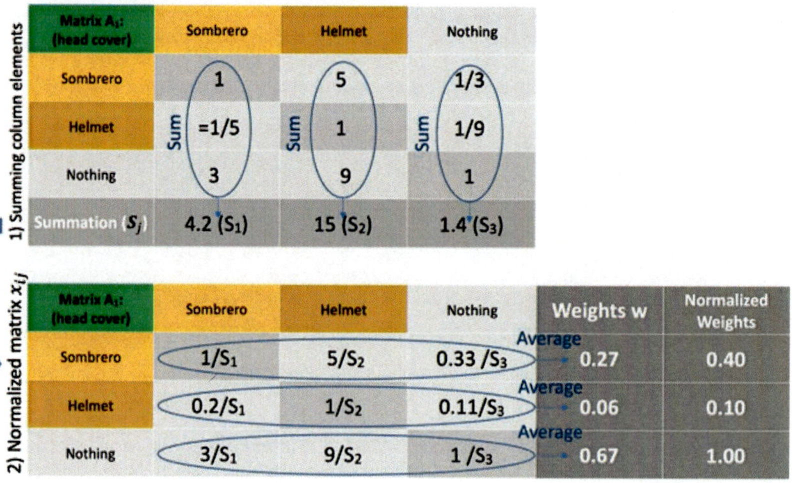

Fig. 2.12 Obtaining normalized (ideal normalization) *Judgment Matrix A_1* and weights by using spreadsheets

> **Note** that the example shown can be easily implemented in a spreadsheet such as Microsoft Excel.

2.7.3 Example: The Consistency Test

In order to evaluate the consistency ratio *CR* in a spreadsheet for the proposed case of the *Vulnerability criterion,* it is necessary to use one more time the matrix A_1 and the weights obtained in the previous step. In particular, the following three steps are necessary:

- First, the columns of the matrix A_1 are *normalized* by using Eq. (2.3).
- Second, it is necessary to evaluate the normalized consistency vector **cv** by using Eq. (2.7). In addition, λ_{max} is estimated by the *average* of the elements of **cv** according to Eq. (2.8).
- Third, CI and CR can be calculated with Eqs. (2.5 and 2.6), respectively.

Figure 2.13 shows the three steps of the procedure in a spreadsheet.

The resulting matrix satisfied the Consistency Ratio requirement CR < 0.1 and hence, the matrix is considered consistent and the weights derived are acceptable.

Figure 2.14 shows how the derived local weights (normalized to 1) can be reported in the flowchart of the structure of the problem. In addition, it is worth noting that the weights are perfectly logical since the most vulnerable situation is represented by the absence of head cover (*Nothing*) with the value 1, the less vulnerable situation is related to the best head cover (*Helmet*) with a *Vulnerability* weight of 0.1. Finally, the *Sombrero hat* represent an intermediate level of *Vulnerability* with a weight of 0.4.

> **Note** that, behind the use of the spreadsheets, there are several software that can support the application of an AHP such as the software "*Expertchoice*".

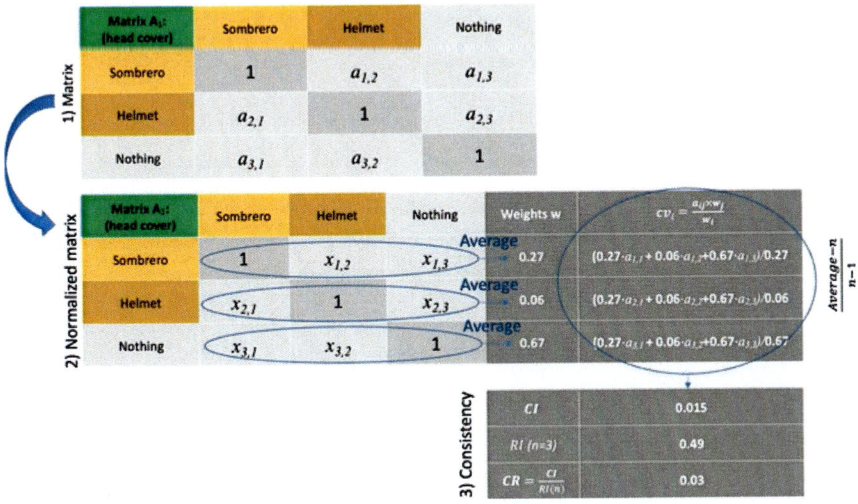

Fig. 2.13 Carrying out the consistency check by using spreadsheets

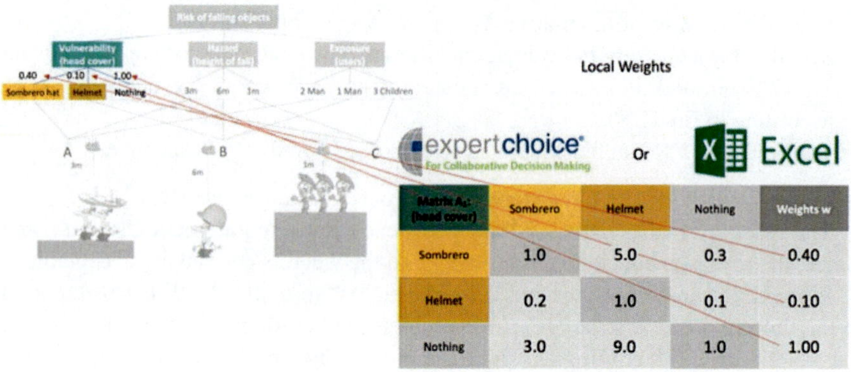

Matrix A_1: (head cover)	Sombrero	Helmet	Nothing	Weights w
Sombrero	1.0	5.0	0.3	0.40
Helmet	0.2	1.0	0.1	0.10
Nothing	3.0	9.0	1.0	1.00

Fig. 2.14 Local weights evaluation and consistency check for the *Vulnerability criterion*

To complete the local weights evaluation, the weights of the alternatives with respect to the *Hazard* and *Exposure* are calculated with the same AHP steps and showed in Figs. 2.15 and 2.16, respectively.

> **Note** that in this example, all the *criteria, Hazard*, *Vulnerability* and *Exposure,* are considered equally important: weight equal to 1. Thus, it is not necessary to evaluate another matrix for the *criteria.*

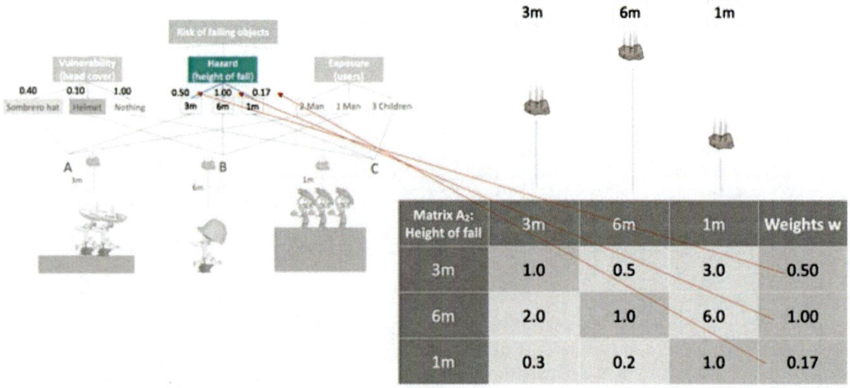

Matrix A_2: Height of fall	3m	6m	1m	Weights w
3m	1.0	0.5	3.0	0.50
6m	2.0	1.0	6.0	1.00
1m	0.3	0.2	1.0	0.17

Fig. 2.15 Local weights evaluation and consistency check for the *Hazard criterion*

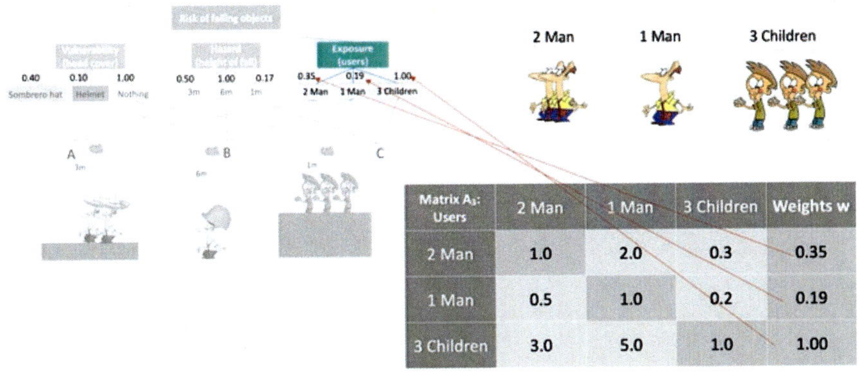

Matrix A_3: Users	2 Man	1 Man	3 Children	Weights w
2 Man	1.0	2.0	0.3	0.35
1 Man	0.5	1.0	0.2	0.19
3 Children	3.0	5.0	1.0	1.00

Fig. 2.16 Local weights evaluation and consistency check for the *Exposure criterion*

2.7.4 *Example: The Global Weights Evaluation in a Simplified Risk Evaluation*

In the example of the "*Risk of falling objects*", once all the local weights are obtained, it is possible to calculate the global weights and the riskiest *alternative* among A, B and C.

At this point, it is important to select the most suitable approach to combine the local weights of *criteria* and *sub-criteria* to achieve the global weights. In this example, the weighted sum (simplest and widespread weights aggregation method existing in literature) is not used because the factorial formula (described in the following) is much more consolidated and applied for the specific case of the risk evaluation (Crichton, 1999). Indeed, in Risk Analysis, based on the *Hazard* (*H*), *Vulnerability* (*V*) and *Exposure* (*E*) dimensions, the global risk (*R*) is typically obtained as the product of the weights related to the three *criteria.* To this aim, the aggregation of the weights is performed as follows:

$$R = V \times H \times E$$

where R is ranging from 0 to 1 (0 represent the null risk and 1 represents the maximum risk).

Once the weights aggregation approach is defined, it is possible to identify the risk of the three *alternatives* (A, B and C) by employing the local weights evaluated as shown in the upper left side of Fig. 2.16 as follows:

$$R_A = 0.40 \times 0.50 \times 0.35 = 0.07$$

$$R_B = 0.10 \times 1 \times 0.19 = 0.02$$

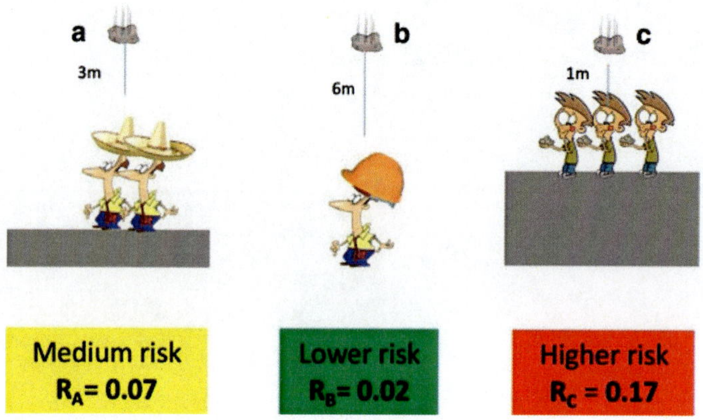

Fig. 2.17 Global weights and ranking of the three *alternatives*

$$R_C = 1.00 \times 0.17 \times 1.00 = 0.17$$

According to this result, the *alternative* A is an intermediate risk situation, the *alternative* B is the lower risk situation and the ***alternative* C** (3 *children* without *head cover* and block falling from 1 m) represent the **higher level of risk**. In conclusion, for this example Fig. 2.17 summarizes the results.

Note that this is just an example and it is not sufficient to make an exhaustive risk analysis in a real case study. In order to obtain a scientific reliable result, it is necessary to *validate* the weights obtained with real data regarding the investigated phenomenon (e.g. analyzing the real injuries regarding the detachment of the building concrete cover). Alternatively, it is possible to compare the results obtained with other similar existing methodologies or set large test sites to confirm the effectiveness of the outcomes.

2.7.5 Example: The Sensitivity Analysis

To verify the robustness of the example results, it is necessary to perform a sensitivity analysis.

The first important preliminary step to perform a *sensitivity analysis* is to understand which part of the decision process should be investigated more in dept and why.

In the proposed application, the global weights' distribution is crucial. On the contrary, the local ranking is not important to be investigated. Therefore, we adopt

Matrix A_1: (head cover)	Sombrero	Helmet	Nothing
Sombrero	1^{μ}	$a_{1,2}{}^{\mu}$	$a_{1,3}{}^{\mu}$
Helmet	$a_{2,1}{}^{\mu}$	1^{μ}	$a_{2,3}{}^{\mu}$
Nothing	$a_{3,1}{}^{\mu}$	$a_{3,2}{}^{\mu}$	1^{μ}

Fig. 2.18 *Sensitivity analysis* on the *Vulnerability criterion*: *Judgment Matrix* $\overline{A_1}$ evaluated by changing μ

the approach proposed by Hurley (2001) for the *sensitivity analysis* which enables the weights to be changed while maintaining the local ranking order. In this method, the focus is on the global weights change and it is useful to evaluate the robustness of the result.

Let us consider a matrix of comparison A having elements a_{ij} and let us define the additional matrices $\overline{A}$, where the elements $\bar{a}_{ij}$ are determined as follows:

$$\overline{a}_{ij} = a_{ij} \tag{2.9}$$

with $\mu \geq 0$. For $\mu - 1$, it holds $A = \overline{A}$; for $\mu > 1$, the resulting weights exhibit the same rank order, but values are "exploded"; for $\mu < 1$, again the resulting weights have the same rank order, but values are "imploded". Such variations in the local weights' variations helps to assess if the stability of the global weights and the overall rank order is preserved. Now, the matrix $\overline{A_1}$ (Fig. 2.18) related to the *Vulnerability* is calculated and weights are obtained by solving (1).

In particular, in this example we let μ vary between 0.8 and 1.2. Figure 2.18 shows $\overline{A_1}$ to illustrate each element a_{ij} of this matrix. For every new matrix ($\mu = 0.8$, $\mu = 0.9$, $\mu = 1.1$ and $\mu = 1.2$) the results can be evaluated by following the simplified procedure of the local weights' evaluation.

Table 2.4 shows the results of the sensitivity analysis. The results show that local weights (weight of *Sombrero*, *Helmet* and *Nothing*) change slightly due to the imposed variation of the *Judgment Matrix*. These minor variations on the weights of Vulnerability produce small variations on the global weights for the risks measures R_A, R_B and R_B.

In addition, the final ranking of the *alternatives* A, B and C remain unchanged for every μ. This important outcome of the *sensitivity analysis* demonstrate that the results are robust and they do not change significantly with changes in the *Vulnerability Judgment Matrix*.

Table 2.4 *Sensitivity analysis* of the *Vulnerability criterion* and relative changes in the final risk indexes

μ	$\mu = 0.8$	$\mu = 0.9$	$\mu = 1$	$\mu = 1.1$	$\mu = 1.2$
Weight of sombrero	0.48	0.44	0.40	0.37	0.30
Weight of helmet	0.15	0.12	0.10	0.08	0.05
Weight of nothing	1.00	1.00	1.00	1.00	1.00
Risk R_A	0.10	0.09	0.08	0.07	0.06
Risk R_B	0.03	0.02	0.02	0.02	0.01
Risk R_C	0.17	0.17	0.17	0.17	0.17

Note that this example has only educational value: a complete *sensitivity analysis* should investigate changes on all the *criteria* or *sub-criteria* and consider all the possible combinations.

References

Brunelli, M. (2014). *Introduction to the analytic hierarchy process.* Springer.

Caporale, D., Sangiorgio, V., Amodio, A., & De Lucia, C. (2020). Multi-criteria and focus group analysis for social acceptance of wind energy. *Energy Policy, 140*, 111387.

Crichton, D. (1999). The Risk triangle. Natural disaster management: a presentation to commemorate the international decade for natural disaster reduction (IDNDR), 1990–2000 Ingleton J: Tudor Rose.

Harker, P. T., & Vargas, L. G. (1987). The theory of ratio scale estimation: Saaty's analytic hierarchy process. *Management Science, 33*(11), 1383–1403.

Hurley, W. J. (2001). The analytic hierarchy process: a note on an approach to sensitivity which preserves rank order. *Computers & Operations Research, 28*(2), 185–188. https://doi.org/10.1016/S0305-0548(99)00125-2.

IEC, I. (2008). ISO 31010—Risk management—Risk assessment techniques. *International Electrotechnical Commission: Geneva.*

Ishizaka, A., & Lusti, M. (2006). How to derive priorities in AHP: A comparative study. *Central European Journal of Operations Research, 14*(4), 387–400.

Miller, G. A. (1994). Reprint of the magical number seven, plus or minus two: Some limits on our capacity for processing information. *Psychological Review, 101*(2), 343–351.

Noble, E. E., & Sanchez, P. P. (1993). A note on the information content of a consistent pairwise comparison judgment matrix of an AHP decision maker. *Theory and Decision, 34*(2), 99–108.

Saaty, T. L. (2003). Decision-making with the AHP: Why is the principal eigenvector necessary. *European Journal of Operational Research, 145*(1), 85–91.

Saaty, T. L. (2008). Decision making with the analytic hierarchy process. *International Journal of Services Sciences, 1*(1), 83–98.

Saaty, T. L., & Vargas, L. G. (1984). Inconsistency and rank preservation. *Journal of Mathematical Psychology, 28*, 205–214.

Saaty, T. L., & Vargas, L. G. (2012a). Models, methods, concepts & applications of the analytic hierarchy process (Vol. 175). Springer Science & Business Media.

Saaty, T. L., & Vargas, L. G. (2012b). The possibility of group choice: Pairwise comparisons and merging functions. *Social Choice and Welfare, 38*(3), 481–496.

Sangiorgio, V., & Parisi, F. (2020). A multicriteria approach for risk assessment of Covid-19 in urban district lockdown. *Safety Science*, 104862.

Sangiorgio, V., Iacobellis, G., Adam, J. M., Uva, G., & Fatiguso, F. (2018). User-Reporting Based Decision Support System for Reinforced Concrete Building Monitoring. In *2018 IEEE international conference on systems, man, and cybernetics (SMC)* (pp. 2620–2625). IEEE.

Sangiorgio, V., Pantoja, J. C., Varum, H., Uva, G., & Fatiguso, F. (2019). Structural degradation assessment of RC buildings: Calibration and comparison of semeiotic-based methodology for decision support system. *Journal of Performance of Constructed Facilities, 33*(2), 04018109.

Sangiorgio, V., Fiorito, F., & Santamouris, M. (2020a). Development of a holistic urban heat island evaluation methodology. *Scientific Reports, 10*(1), 1–13.

Sangiorgio, V., Uva, G., & Aiello, M. A. (2020b). A multi-criteria-based procedure for the robust definition of algorithms aimed at fast seismic risk assessment of existing RC buildings. In *Structures* (Vol. 24, pp. 766–782). Elsevier.

Towhata, I., (2008). *Geotechnical earthquake engineering* by, ed. Springer. https://doi.org/10.1007/978-3-540-35783-4.

Sangiorgio, V., Martiradonna, S., Fatiguso, F., & Lombillo, I. (2021). Augmented reality based-decision making (AR-DM) to support multi-criteria analysis in constructions. *Automation in Construction, 124*, 103567. https://doi.org/10.1016/j.autcon.2021.103567.

Vargas, L. G. (2008). The consistency index in reciprocal matrices: comparison of deterministic and statistical approaches. *European Journal of Operational Research,* 191(2), 454–463. https://doi.org/10.1016/j.ejor.2007.06.054.

Chapter 3
Augmented Reality to Support the Analytic Hierarchy Process

Abstract In the era of "Digital Transition", also traditional and consolidated approaches such as the *Analytic Hierarchy Process* (AHP) can benefit from technological innovations. The AR-AHP is particularly effective since the DM can be supported by 3D models displayed in AR and a simple weights evaluation procedure. Indeed, the *Augmented Reality* (AR) is particularly effective to support decision analysis in the field of **architectural engineering** and **building production** by showing 3D models (specifically developed) directly on-site. Indeed, technical drawing, axonometry representation and three-dimensional models have been used for centuries for their ability to visually transmit useful information to the user. In this modern concept, the visual support of technical drawing achieves greater effectiveness through the AR environment. This chapter shows how augmented reality can support the *Analytic Hierarchy Process* (AR-AHP) to get a powerful multi-criteria approach. The AR-AHP exploits an AR environment to visually provide a large amount of information during the decision process. In particular, this approach exploits the *structuring of the problem* analogously to the AHP to decompose the decision problem into independent criteria and transform a multidimensional scaling problem into many one-dimensional scaling problems. Afterwards, the *local weights evaluation* exploits the AR technology and an adapted version of the Simos-Roy-Figueira (SRF) method for an effective extraction of weights (entirely performed in the AR environment).

Keywords Augmented reality · Analytic hierarchy process · Architectural engineering · Building production · Technological innovation · Simos-Roy-Figueira method · Building 3D modelling

In the era of "Digital Transition" of the building sector, also traditional and consolidated approaches such as the AHP can benefit from technological innovations. This chapter shows how Augmented Reality can support Decision Making (AR-AHP) to get a powerful multi-criteria decision approach (Sangiorgio et al., 2021). The AR-AHP exploits the augmented environment to visually provide a large amount of information during the decision process. This approach exploits the *structuring of the problem* analogously to the AHP (Saaty, 2008) to decompose the decision problem into independent criteria and transform a multidimensional scaling problem

V. Sangiorgio et al., *New Approaches for Multi-Criteria Analysis in Building Constructions*, https://doi.org/10.1007/978-3-030-83875-1_3

into many one-dimensional scaling problems. Consequently, also the synthesis of priority (*global weights evaluation*) follows the classical AHP in order to aggregate weights of all of the many one-dimensional scaling problems. On the contrary, the *local weights evaluation* exploits the AR technology and an adapted version of the Simos-Roy-Figueira (SRF) method (Figueira & Roy, 2002) for an effective weight's extraction. Indeed, the SRF approach is very effective to be adapted and used in combination with AR. In this way, the visual information provided in the AR allows effectively carrying out a complex decision even if the DM does not have an exhaustive knowledge of the problem.

The novelties of the proposed method are listed in the following items:

(i) Compared to AHP, in the AR-AHP the DM can perform the comparison of the involved parameters exploiting the SRF procedure in a useful AR environment. Moreover, even non-expert users of the specific problem can successfully carry out the procedure thanks to the useful information displayed in the virtual 3D models.
(ii) Compared to SRF procedure, the weights evaluation procedure is applied several times to obtain all the weights defined in the hierarchical structuring of the problem according to the AHP. In addition, the algorithm for weights evaluation is improved, based on the aggregation principles of local weights of the AHP.
(iii) Adapted consistency tests can be used to assess the reliability of the local weights.

3.1 Overview of the Method: The Six Steps of the AR-AHP

In this section, the *6 steps* of the *AR-AHP* procedure are presented and a flowchart (Fig. 3.1) shows the connection and interactions among the *steps*. In particular, AR-AHP methodology is composed of the following steps:

AR-AHP ***Step (1)*** the *Structuring the problem* that follows the same theory of the classical AHP;
AR-AHP ***Step (2)*** the *AR setting* devoted to creating the "*Augmented Reality* environment";
AR-AHP ***Step (3)*** the *Local weights evaluation* in an AR environment. Such evaluation is performed by exploiting the 3D models in augmented reality according to the following three sub-steps named "*AR phases*":

AR phase 1—Local preliminary ranking in which the DM orders the 3D models to perform a local preliminary ranking in the augmented reality environment,
AR phase 2—Local ranking in which the DM compares 3D models in pairs and achieve the local ranking,
AR phase 3—Local weights evaluation performed by exploiting the SRF theory;

Fig. 3.1 Flowchart of the AR-AHP

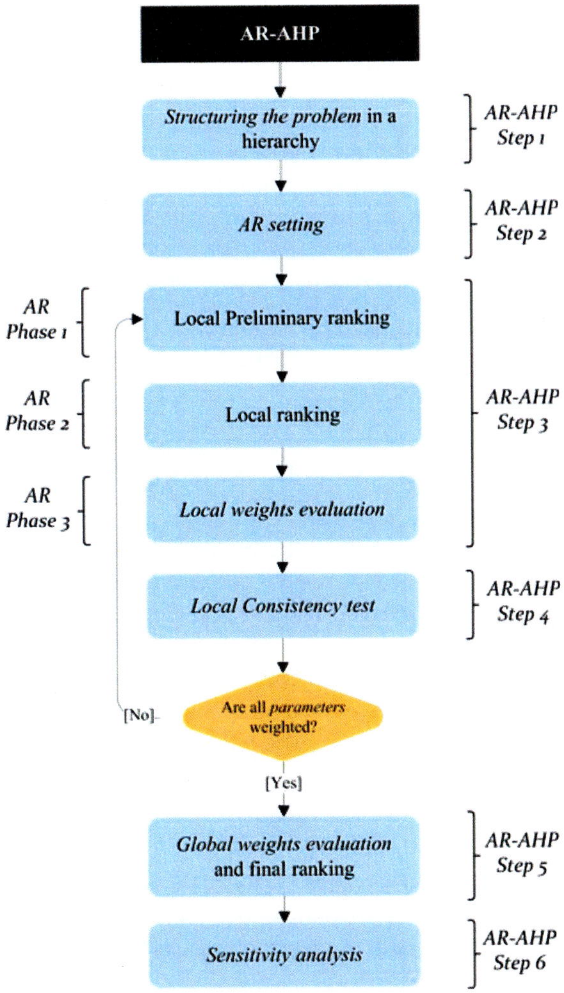

AR-AHP **Step (4)** a specific *Local consistency test* to verify the coherence of the *Augmented Reality* supported weighting;
AR-AHP **Step (5)** the *Global weights evaluation* that can be obtained with different existing formulations (depending on the problem investigated) as for the classical AHP;
AR-AHP **Step (6)** the *Sensitivity analysis* that like classical AHP can be focused on different parameters.

Note that *Steps 3* and *4* are repeated until all the *parameters* are weighted.

3.2 AR-AHP Step 1: *Structuring the Problem*

Step 1 follows the footstep of the AHP (Saaty, 2008). This first stage is devoted to decomposing and structuring the problem in a flowchart to obtain a complete overview of the involved *macro-criteria, criteria, sub-criteria* or *alternatives* according to the classical procedure (Sect. 2.1). The goal of the analytical process is stated in the initial part of the flowchart. Successively the involved parameters which may include *macro-criteria, criteria, sub-criteria* and *alternatives* are classified in different levels, to reach a complete description of the considered problem.

Remember that the term *parameters* is used to generically refer to *macro-criteria, criteria, sub-criteria,* or *alternatives.*

3.3 AR-AHP Step 2: *AR Setting* and 3D Models Design

In *AR-AHP step 2* the AR environment and the virtual 3D models are set for the analysis. This step is essential to create useful information displayable in augmented reality. In the virtual 3D models, effective visual content can be included to support the DMs to face the analysis. In particular, a 3D model can be designed and customized for every parameter. Indeed, a 3D model can represent a specific *parameter* in relation to a specific *criterion* (*sub-criterion* or *macro-criterion*) and may include useful visual information connected to the *criterion.*

For example, every 3D model could enclose (but not only) the following information:

(i) The name of the *parameter*;
(ii) The representative 3D model or a graphical scheme;
(iii) A useful description or qualitative information connected to the *parameter*;
(iv) Quantitative data, which can be typological, functional, characteristic or economic;
(v) Advantages; and
(vi) Aisadvantages or other information.

Note that the number of 3D models that necessary depend on the number and typology of *parameters* to be analyzed and structured in the flowchart of *AR-AHP Step 1.* It is not necessary that all the *parameters* are represented by 3D models since some parameters can also be quantitative (e.g. time, costs, specific performance).

3.4 AR-AHP Step 3: Evaluation of *Local Weights*

After all the 3D models have been created, the operational part of the *weights evaluation* procedure can start.

The DM is asked to perform the *local weights evaluation* by analyzing every single aspect of the decision problem in the AR environment. In particular, the AR supported analysis is carried out in three phases named *AR Phase 1*, *AR Phase 2* and *AR Phase 3*. These three *AR* phases are developed by following the footsteps of the SRF theory according to (Sangiorgio et al., 2021) and are explained as follows.

3.4.1 AR Phase 1—Local Preliminary Ranking

Let us assume that a set of 3D models is defined to evaluate a set of *parameters*.

In *AR Phase 1* the user is asked to rank the 3D models (representing a set of *parameters* in relation to a specific *criterion*) from the less important to the most important. So, the user orders the 3D models in an AR environment by assigning to them ascending importance to obtain a preliminary local ranking: using a one-dimensional space and positioning the models from left to right in order of importance, the 3D models positioned on the left correspond to an assignment of low importance and hence, 3D models positioned on the right part of the AR environment are the most important. In addition, if the DM decides that some 3D models have the same importance (i.e. the same weight), the DM can assign the same position to a subset of 3D models. Consequently, the output of *AR Phase 1* is a *Local preliminary ranking* of the considered parameters.

Figure 3.2 shows five 3D models showing a sketch of five different buildings and a general example of a ranking of the importance of these constructions. In this example, building #1 is positioned to the left (lower importance). Building #2 is

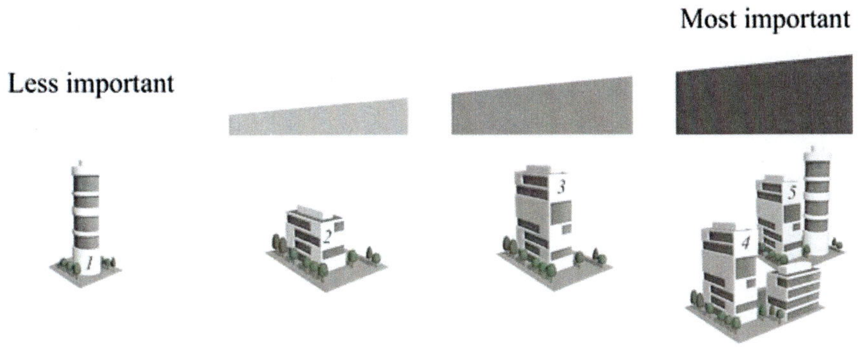

Fig. 3.2 Example of AR phase 1: local preliminary ranking of a set of 3D models (representing parameters)

positioned to the right of building #1 and in turn building #3 is positioned to the right of Building #2. Buildings number #4 and #5 are the most important and have the same significance. Thus, these 3D models are assigned to the same position to the right. In this phase, the 3D models are arranged in a line at equally spaced intervals.

3.4.2 AR Phase 2—Local Ranking

In *AR Phase 2*, the user can decide how large the differences of two successive 3D models (or subsets of 3D models) are. This operation is made by comparing in pairs adjacent 3D models. The users can introduce empty spaces (represented with white cubes) between two successive 3D models (or subsets of 3D models) in order to increase their differences in relation to the local preliminary ranking. The absence of empty spaces between two consecutive 3D models means a small difference. The presence of a large number of empty spaces between two consecutive 3D models means a large difference between the parameters (represented with the 3D models). The *AR Phase 2* finishes with the assignment of a rank (*local ranking*), both to the 3D models, empty spaces and subsets of 3D models (Fig. 3.3).

Figure 3.3 shows in more detail the *local ranking* obtained after the introduction of three empty spaces. For example, Fig. 3.3 shows that if building #2 is slightly more important than building #1, then no empty spaces need to be positioned between them. On the contrary, if the difference between building #2 and #3 is moderate, a single empty space can be included between them. Moreover, the presence of a large difference can justify inserting two or more empty spaces as it happens between building #3 and the group of buildings #4 and #5. Once all the necessary empty spaces are included, the *local ranking* can be obtained and every position of the ranking is correlated with a natural number starting from the less important (for which assigned the value 1) to the most important. In the proposed example, group buildings #4 and #5 have the highest value and they are assigned the value 7.

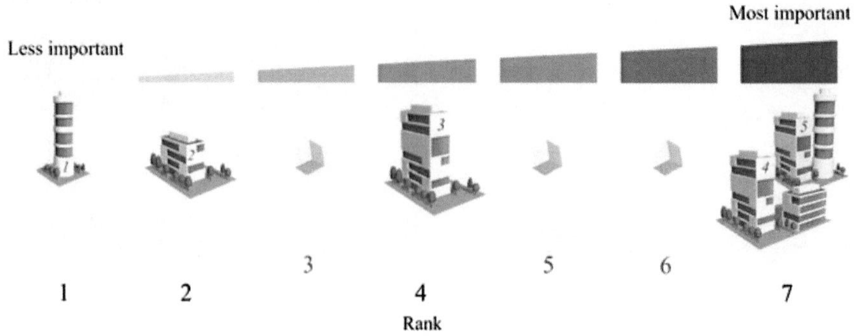

Fig. 3.3 Example of *AR Phase 2*: *local ranking* obtained using 3D models and empty spaces (white cubes)

3.4.3 AR Phase 3—Local Weights Evaluation

In the *AR Phase 3* it is possible to derive weights from the result of *AR Phase 2.*

In particular, the local weights' evaluation of the *AR Phase 3* can be achieved as described in the following.

First, starting from the *local ranking*, it is necessary to consider the assigned rank and related natural number starting from the less important to the most important consideration. It is worth noting that, the rank is assigned both to the model and empty spaces starting from 1 (assigned to the less important 3D model), proceeding with 2, 3, etc. (assigned to any 3D model or empty space) until the most important 3D model.

Second, the ranks assigned to the 3D models (only 3D models and not considering the empty spaces) are normalized to 1 (every rank is divided for the sum of the rank of the 3D models) to obtain the weights (Fig. 3.4).

Note that this procedure is developed and validated in Sangiorgio et al. (2021) starting from the theory of SRF. Compared with the classical SRF method, the algorithm for weights evaluation is improved, based on the aggregation principles of local weights of the AHP in order to apply the weights evaluation procedure several times and weight all the considered parameters.

In what follows, the *AR Phase 3* described above is mathematically formalized according to Sangiorgio et al. (2021) (by defining parameters, 3D models, empty

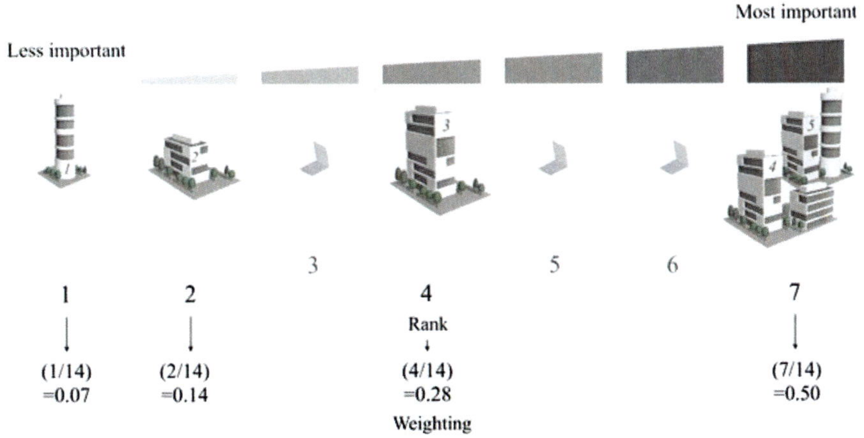

Fig. 3.4 Example of *AR phase 3—local weights evaluation*: local ranking obtained using 3D models and empty spaces (white cubes)

spaces and ranks) to extract weights from the local ranking obtained by using 3D models and empty spaces.

Let us assume that a set $N = \{p|p = 1, \ldots, n\}$ of parameters (with $n \in \mathbb{N}$ where $\mathbb{N}$ is the set of natural numbers) are analyzed through the *AR phases* and let us denote $C = \{c_p|p = 1, \ldots, n\}$ the set of 3D models c_p, where c_p is associated to the *p*-th parameter for $p = 1, \ldots, n$. In addition, we define the set $E = \{q|q = 0, \ldots, m\}$ (with $m \in \mathbb{N}$) of empty spaces. A rank $r_p \in \{1, \ldots, n + m\}$ is assigned to each 3D model $c_p \in$ C during *AR Phase 2.* The local weight $v_p \in \mathbb{R}^+$ associated to each 3D model $c_p \in C$ (and then to each parameter *p)* is given by:

$$v_p = \frac{r_p}{\sum_{p=1}^{n} r_p} \quad \text{for } p = 1, \ldots, n \tag{3.1}$$

where $\mathbb{R}^+$ is the set of l positive real numbers. Note that $v_p \in [0, 1]$ with $\sum_{p=1}^{n} v_p = 1$. We remark that Eq. (3.1) is defined in order to respect the normalization principle used in the extraction of weights from Saaty's positive reciprocal matrices obtained by considering the perception of DMs where the weights are normalized to 1.

Note that analogously to AHP, the proposed AR-AHP can include also quantitative parameters in the analysis. If the DM needs to evaluate a set of *parameters N* composed by numerical values of the same unit of measurement, the weights of these *parameters* can be directly obtained normalizing the numerical values of the parameters. This possibility is showed in a simplified example (Sect. 3.7.4).

3.5 AR-AHP Step 4: *Local Consistency Test*

In addition, a suitable local consistency test can be performed in order to verify whether the user is aware of the choices made in the *AR phases,* and to verify that the resulting weighs are coherent. This test is inspired by the redundant judgments matrices of the AHP. Indeed, analogously with the AHP, the AR-AHP consistency test is based on additional and redundant judgments to verify the coherence of assigned rankings and empty spaces (Sangiorgio et al., 2021). In addition, this test is devised to be compatible with the proposed *AR phase 2.*

In particular, in the AR system the DM is asked to perform an additional 3D model pair comparison (the 3D models are randomly extracted from the set *C*) expressing numerically how much one parameter (associated with the 3D models) is more important than the other ones. Let us assume that the user extracts the 3D models c_i and c_j and assigns a value *k* to this paired comparison. The value *k* represents the differences between c_i and c_j according to the DM's perception. In addition, if the

DM has consistently applied the local and global ranking of *AR Phase 2,* then *k* should have a similar value to the ratio between v_i and v_j (that are the weights obtained in *AR Phase 2*).

Consequently, the local consistency between c_i and c_j, denoted LC(*i, j*), is evaluated by the following formula:

$$\text{LC}(i,\ j) = \left| \frac{\text{k} - \left(v_i / v_j\right)}{v_i / v_j} \right| \quad \text{for each } c_i, c_j \in C \quad \text{with } i \neq j \tag{3.2}$$

On the basis of several empirical tests in similarity with Saaty (2008), the work of Sangiorgio et al. (2021) assumes that the values LC(*i,j*) < 0.30 are acceptable.

Note that this first local consistency test does not ensure that the final result is reliable, but it indicates whether the user has a good perception of the importance of the analyzed parameters.

3.6 AR-AHP Step 5: The Global Weights Evaluation

In the *AR-AHP step 5*, the *global weights evaluation (*synthesis of priority or weights aggregation) is performed to determine the rankings and the global weights for each *alternative* by following the same procedure of the classical AHP. To this aim, the weights of *criteria* are combined with the weights of the *alternatives* in order to obtain the global weights (Sect. 2.4).

Remember that in the related literature there are many equations to perform the weights aggregation and the use of a specific equation depends on the problem investigated.

3.7 Example: AR-AHP Application for Simplified Risk Evaluation in the Field of Building Construction Management

In order to present a familiar example and allow a comparison with the AHP, the example of the risk for concrete blocks falling down is proposed again in this chapter.

3.7.1 Example: AR-AHP Step 1, Structuring the Problem

Remember that the ***objective*** of the example can be summarized in the evaluation of the "***Risk of falling objects***" and the three ***alternatives*** (Fig. 2.8, Sect. 2.7.1) are: (A) *two man* with *sombrero hat* and block falling from *3 m*; (B) *one man* with *helmet* and block falling from 6 *m*; and (C) *three children* without head cover (*nothing*) and block falling from 1 *m*.

3.7.2 Example: AR-AHP Step 2, AR Setting *and 3D Models Design*

The AR-AHP Step 2 allows including information in augmented reality during the decision phases. Consequently, the DM can exploit the visual information obtainable by rotating and displaying the 3D model together with additional characteristics viewable in AR. The proposed example shows how some useful information in the 3D model of the man with *sombrero hat* can be included (Fig. 3.5).

> **Note** that this example is very simple but a more complex application of the AR-AHP with related complex 3D models is proposed in detail in Chap. 5.

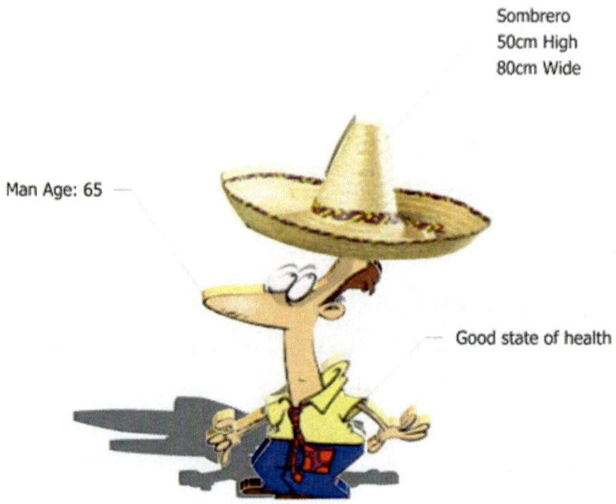

Fig. 3.5 Example of the introduction of useful information in a 3D model

3.7.3 *Example: AR-AHP Step 3,* Local Weights Evaluation

In Step 3, Starting from the first *criterion* of *Vulnerability*, the DM can obtain the local weights of the three sub-criteria *Sombrero hat, Helmet or Nothing* by performing the three *AR phases.*

In the *AR phase 1* the first *local preliminary ranking* can be performed. The DM is asked to rank the three head covers from the less vulnerable to the much vulnerable. Evidently, *Helmet* is the less vulnerable, *nothing* is the most vulnerable and *Sombrero* is in the middle position among the *sub-criteria.* The *local preliminary ranking* of the *sub-criteria* with respect to the *criterion* of *Vulnerability* obtained with the 3D models is showed in Fig. 3.6.

The *AR phase 2* yields the *local ranking*. The DM is asked to include empty spaces to improve the difference between the 3D models (parameters). In this example, the DM includes two empty spaces between *Helmet* and *Sombrero* and five empty spaces between *Sombrero* and *Nothing* according to his experience (Fig. 3.7). In addition, note the rank assigned to the 3D models (*Helmet* $r_p = 1$, *Sombrero* $r_p = 4$ and *Nothing* $r_p = 10$).

In the *AR phase 3* the *local weights* can be evaluated by using Eq. (3.1) by considering the rank assigned to the 3D models. In addition, a simple spreadsheet can be used to perform the weight evaluation. Figure 3.8 shows the weights' evaluation of the *sub-criteria* stating from the rank assigned to the corresponding 3D model exploiting Eq. (3.1).

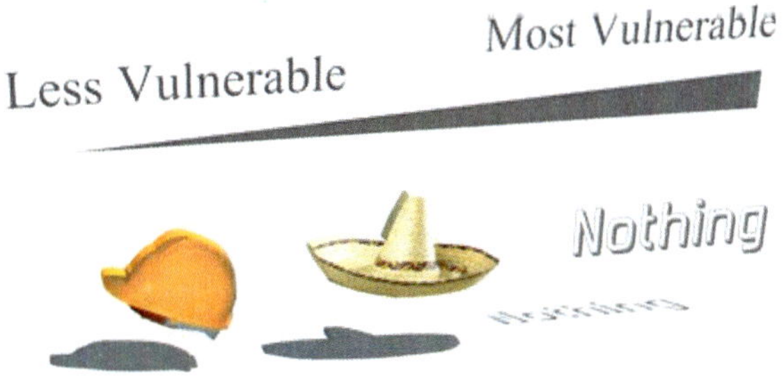

Fig. 3.6 *Local preliminary ranking* of *sub-criteria* with respect to the *criterion* of *Vulnerability*

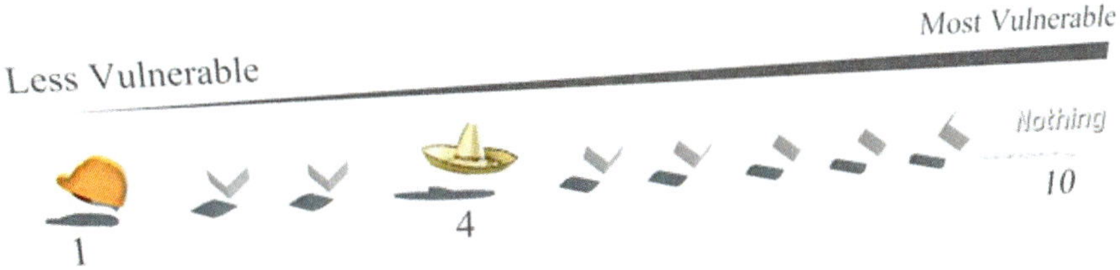

Fig. 3.7 *Local ranking* of *sub-criteria* with respect to the *criterion* of *Vulnerability*

Criterion	*Sub-criteria*	*Ranking*	$\frac{r_p}{\sum_{p=1}^{n} r_p}$ (10)	*Weights* $\boldsymbol{w}$	*Normalized to 1*
	Sombrero	4	4/(4+1+10)	0.27	0.40
Vulnerability	*Helmet*	1	1/(4+1+10)	0.07	0.10
	Nothing	10	10/(4+1+10)	0.67	1.00

Fig. 3.8 *Local weights* evaluation of *sub-criteria* with respect to the *criterion* of *Vulnerability*

Note that the weight obtained in this example by using the AR-AHP approach is exactly the same as the weight obtained by using the classical AHP procedure in Sect. 2.7.3.

3.7.4 Example: AR-AHP Step 4, Local Consistency Test

At this point, it is necessary to perform the *Local consistency test*. The DM is asked to perform an additional 3D model pair comparison (the 3D models are randomly extracted from the set *Helmet, Sombrero,* and *Nothing*). The extracted parameters are *Sombrero* and the *Helmet*. Indeed, the DM is asked according to his perception to evaluate how much (numerically, e.g. on a scale from 1 to 10) the *Sombrero* is more *vulnerable* of the *Helmet*.

Figure 3.9 shows an example where the DM answer is that *Sombrero* are 4 times more vulnerable than *Helmet* ($k = 4$).

By reworking Eq. (3.2), it is possible to verify the Local Consistency.

$$\text{LC(Sombrero, Helmet)} = \left| \frac{k - (w_{\text{Sombrero}}/w_{\text{Helmet}})}{w_{\text{Sombrero}}/w_{\text{Helmet}}} \right| = \left| \frac{4 - (0.27/0.07)}{0.27/0.07} \right| \cong 0.03$$

Consequently, the LC *(Sombrero, Helmet)* < 0.30 is acceptable.

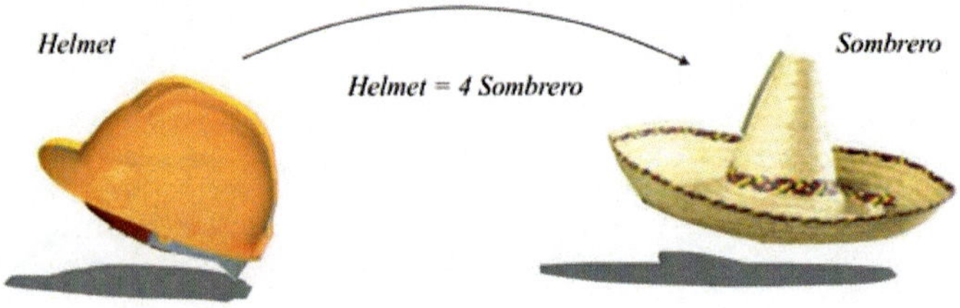

Fig. 3.9 Example of the *local consistency test* (two *sub-criteria* with respect to the *criterion* of *Vulnerability*)

Criterion	*Sub-criteria*	*Quantitative (m)*	$\frac{r_p}{\sum_{p=1}^{n} r_p}$	*Weights* $\boldsymbol{w}$	*Normalized to 1*
	3m	3	3/(3+6+1)	0.30	0.50
Hazard	*6m*	6	6/(3+6+1)	0.60	1.00
	1m	1	1/(3+6+1)	0.10	0.17

Fig. 3.10 Quantitative *local weights* evaluation of a *sub-criteria* with respect to the *criterion* of *Vulnerability*

After the consistency of the firsts *sub-criteria* with respect to the *criterion* of *Vulnerability* is verified, it is necessary to evaluate the weighs the *sub-criteria* in relation to the other *criteria* before proceeding with the last AR-AHP step.

Considering the *criterion Hazard*, the weights of the three *sub-criteria, 3 m, 6 m* or *1 m* can be extracted quantitatively analogously to AHP. The weights of this *parameters* can be directly obtained normalizing the numerical values of the parameters as showed in Fig. 3.10.

Note that, also in this case the weights are the same obtained by using the classical AHP in the Sect. 2.7.3.

Considering the criterion *Exposure*, the weights of the three *sub-criteria* (*two men, one man* and *three children*) can be evaluated by repeating the three *AR phases*. The results of the three phases are shown in Figs. 3.11, 3.12 and 3.13, respectively.

First, the *local preliminary ranking* is obtained identifying the *one man* as the less exposed, the *two men* in the intermediate position and the *three children* as the most exposed among the *sub-criteria* (Fig. 3.11).

Second, the *local ranking* is obtained by including empty spaces. No empty spaces are included between *one man* and *two men* (meaning a small difference in exposure) and two empty spaces are inserted between *two men* and *three children* (meaning a medium/high difference in exposure) (Fig. 3.12).

Finally, the *local weights* evaluation of the *sub-criteria* with respect to the *criterion* of *Exposure* is achieved using the *local ranking* and Eq. (3.1) (Fig. 3.13).

Fig. 3.11 *Local preliminary ranking* of *sub-criteria* with respect to the *criterion* of *exposure*

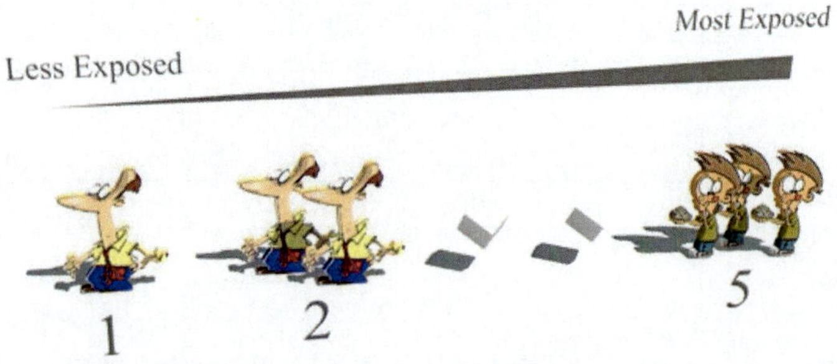

Fig. 3.12 *Local ranking* of *sub-criteria* with respect to the *criterion* of *exposure*

Criterion	*Sub-criteria*	*Ranking*	$\frac{r_p}{\sum_{p=1}^{n} r_p}$ (10)	*Weights* $\boldsymbol{w}$	*Normalized to 1*
	2 Man	2	2/(2+1+5)	0.25	0.40
Exposure	*1 Man*	1	1/(2+1+5)	0.13	0.20
	3 Children	5	5/(2+1+5)	0.63	1.00

Fig. 3.13 *Local weights* evaluation of *sub-criteria* with respect to the *criterion* of *exposure*

3.7.5 Example: AR-AHP Step 5, the Global Weights Evaluation

After obtaining all the *local weights* of the problem, they can be stored and presented as tabulated weights (Fig. 3.14).

Finally, the global weighs can be obtained by using the factorial formula analogously to the example of Chap. 2.

Fig. 3.14 *Local weights* stored and presented as tabulated weights

Criterion	*Sub-criteria*	*Normalized to 1*
	Sombrero	0.40
Vulnerability	*Helmet*	0.10
	Nothing	1.00
	3m	0.50
Hazard	*6m*	1.00
	1m	0.17
	2 Man	0.40
Exposure	*1 Man*	0.20
	3 Children	1.00

Note that all the local weights are almost exactly the same as the weights found in the example performed by the AHP (Sect. 2.7.4). Consequently, the global weighs evaluation is analogous to the example of the AHP, reported as follows:

$$R_A = 0.4 \times 0.5 \times 0.4 = 0.08.$$

$$R_B = 0.1 \times 1 \times 0.2 = 0.02.$$

$$R_C = 1.0 \times 0.17 \times 1 = 0.17.$$

References

Figueira, J., & Roy, B. (2002). Determining the weights of criteria in the ELECTRE type methods with a revised Simos' procedure. *European Journal of Operational Research, 139*(2), 317–326.

Saaty, T. L. (2008). Decision making with the analytic hierarchy process. *International Journal of Services Sciences, 1*(1), 83–98.

Sangiorgio, V., Martiradonna, S., Fatiguso, F., & Lombillo, I. (2021). Augmented reality based-decision making (AR-DM) to support multi-criteria analysis in constructions. *Automation in Construction, 124*, 103567.

Chapter 4
How to Set a User Reporting Supported Decision Making in Architectural Engineering and Building Production

Abstract The importance of acquiring information from users is widely recognized by the scientific and technical community in the building sector. In addition, the advent of novel technologies (e.g. smart devices) offers unprecedented observational capacity at the scale of the individual. This chapter describes the general framework, methodology and technologies to set up a *User Reporting* supported decision analysis, useful in all the phases of the building process. In particular, the *User Reporting* could be used for two different along with multi-criteria analysis (MCA): (i) support the weighting in an MCA by involving a large set of decision makers; (ii) apply performance indexes resulting from an MCA (previously developed) on a large scale. In particular, the *User Reporting* is explained both from the theoretical and practical point of view and consists of the following eight steps:

(1) *Identification of the stakeholders*;
(2) *Definition of the Users*;
(3) *Selection of technological tools* to support the acquisition;
(4) *Creation of questionnaires and guidelines*;
(5) *Definition of the flowchart;*
(6) *Effective Data acquisition*;
(7) *Data processing and validation*;
(8) *Data analysis.*

A simplified example clarifies the eight steps by considering the risk analysis of a heavy expulsion of concrete cover in residential buildings.

Keywords User Reporting · Architectural engineering · Building production · Technological tools · Stakeholders analysis · Questionnaires · Participatory sensing · Data analysis

In the building sector, the possibility of acquiring information directly by the user can be useful in all the phases of the building process *design*, *construction*, *management* and *dismantling*. In addition, the advent of novel technologies (e.g. smart devices) offers unprecedented observational capacity at the scale of the individual.

V. Sangiorgio et al., *New Approaches for Multi-Criteria Analysis in Building Constructions*, https://doi.org/10.1007/978-3-030-83875-1_4

This section describes the general framework, methodology and technologies to set up a *User Reporting* supported decision analysis in construction.

User Reporting regards the methodological approach aimed at obtaining relevant information directly from the users of the buildings by analyzing: (i) the users' perception; (ii) direct on-site measurement detected directly by the users (both qualitative and quantitative) following specific guided procedures; (iii) other specific measurements acquired by the users exploiting the increasingly high-performance and technological smart devices (e.g. photo, video, audio, temperature, accelerometer registration) (Sangiorgio et al., 2019a, 2019b, 2020). The information that the users can acquire, collect and provide to the engineers and technicians are called *reports*. *User Reporting* can be really effective when supported by technological tools (mobile devices and related application software—APP) in order to include additional useful information in the *reports* (Sangiorgio et al., 2018a, 2018b). For example, the use of mobile devices can enable acquisition of data such as: photographic images, intensity of noise, geolocation, distance, speed, accelerations, vibrations and other specific information obtained with a guided questionnaire.

User Reporting could be used for different purposes along with the MCA. Hence, the first step is to identify the purposes of *User Reporting* between the following two cases:

First Case: "*User Reporting supported decision making*"—this approach can be used to support the weighting in an MCA by considering a large set of users (in this case users are also the DMs of the MCA). In this purpose, *User Reporting* is used to allow a very wide application of the MCA being able to involve hundreds of DMs in the weight evaluation.

Second Case: "*User Reporting supported large-scale analysis*"—*User Reporting* is used to apply KPIs or specific Condition Ratings resulting from an MCA at *t* large scale. In such purpose, the MCA is previously developed and the related result can be used to achieve concise indexes (KPIs or Condition Ratings) which are applied on a large scale through a *User Reporting* supported data acquisition.

Whatever the purpose of the *User Reporting* is, it is fundamental to define the purpose of *User Reporting,* the necessary information to be acquired and eventually the specific MCA that can be included in the approach. This is termed the **preliminary** step.

User-Reporting has eight steps:

(1) *Identification of the stakeholders*;
(2) *Definition of the Users*;
(3) *Selection of technological tools* to support the acquisition (e.g. Smart Devices);
(4) *Creation of questionnaires and user guidelines* (or any choice experiments);
(5) *Definition of the flowchart* (describing all the User Reporting processes);
(6) *Effective Data acquisition*;
(7) *Data processing and validation*; and
(8) *Data analysis.*

Note that the first five steps are developed to set the procedure, starting from the sixth step the effective *Data Acquisition* begins and in the last two steps *Data Processing* and *Data Analysis* are performed. Data processing and analysis are performed according to the defined *User Reporting* **purpose** and exploiting any MCA included in the approach.

In addition, in the final section of this chapter, a simplified example is proposed by considering again the example of the heavy expulsion of concrete cover and rebars oxidation in the residential building sector (the same example given in Chap. 2). More complex applications are proposed in the next chapters for both the two *User Reporting* purposes: to support the weighting in an MCA regarding the social acceptance of wind energy (**First Case**—Chap. 6) and get a large test site supporting building maintenance and diagnostics in the region of Valencia, Spain (**Second Case**—Chap. 7).

4.1 Identification of the Stakeholders

The stakeholders are individuals or groups of people who have an interest or some kind of interest in the problem at hand.

The identification of stakeholders typically starts with brainstorming. The participants in the brainstorm are the *Decision Managers* (researchers or practitioners) interesting in set the *User Reporting* and perform MCA analysis (these people could also be identified as stakeholders or can be the DMs of the MCA).

Note that the *Decision Managers* are the figures who deal with the development of the decision problem and obtaining the results of the analysis. On the other hand, the DMs are the people involved in the MCA that assign the judgments (may or may not correspond with the *Decision Managers*).

By focusing on the *User Reporting* purpose, the *Decision Managers* gather a list of potential stakeholders involved in the decision process. It is important to identify all the people who are involved in the decision problem and in the *User Reporting*, who have influence or power over it, or have an interest in its successful or unsuccessful conclusion. All the potential stakeholders are noted down without criticism and after the brainstorming session, the ideas are evaluated.

The brainstorming can also be performed later (in step 2 of *Definition of Users*), but it is necessary to identify all the potentially crucial stakeholders at this early stage.

The list of potential stakeholders could include mayors or politicians, administrators, different typologies of construction users, constructions staff, product, development/engineering/manufacturing, services, companies, consultants, operations/IT and so on.

Once the stakeholders are identified, it is necessary to classify them into different groups. A *Power/Interest Grid* is a powerful support tool to map out the stakeholders and classify them according to their power over the work and their interest in it (Fig. 4.1). The position of a stakeholder on the grid influences the actions that are necessary to be taken with them:

- **High power, highly interested people (Manage Closely)**: it is important to fully engage these people and make the greatest efforts to satisfy them.
- **High power, less interested people (Keep Satisfied)**: it could be useful to put enough work in with these people to keep them satisfied, but not so much that they become bored with the proposed message.
- **Low power, highly interested people (Keep Informed)**: these people should be adequately informed and their opinion need to be considered to ensure that no major issues arise. People in this category may often be very helpful with the details of the project.
- **Low power, less interested people (Monitor)**: again, these people can be monitored, but it is not recommended to provide them with excessive information.

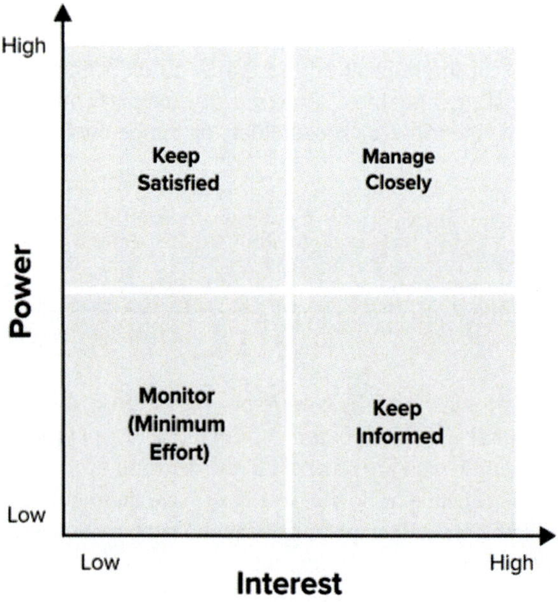

Fig. 4.1 *Power/interest grid* to support stakeholder's classification

4.2 Definition of Users

The *users* are individuals or groups of people which are identified among the stakeholders for their characteristics of being able to provide useful information (good quantity and quality) during *User Reporting*.

Consequently, once the stakeholders are specified, it is fundamental to identify the *users* among them. To this aim, the stakeholders are classified on the basis of the typology and quality of information they can provide during *User Reporting*. In this step, the *users* of the procedure are defined among the identified stakeholders. A correct classification of both *users* and remaining stakeholders (based on different levels of knowledge, quality and quantity of reportable data) is crucial to effectively carry out the successive steps of *User Reporting*. Also in this step, a brainstorming session could be used to explore different ideas and obtain an effective classification of stakeholders. In addition, we propose an adapted version of the *Power/Interest grid* to create and classify *users* and stakeholders according to their levels of knowledge and numerosity named *Numerosity/Expertise grid*.

Note that "knowledge" is a generic term. In this case, the knowledge referred to is the defined purpose of the *User Reporting* and the necessary information to be acquired as it is stated in the preliminary stage. Remember that in the "**preliminary step** it is fundamental to define the purpose of *User Reporting*, identify the necessary information to be acquired and eventually the specific MCA that can be included in the approach."

This novel grid, named *Numerosity/Expertise grid* (Fig. 4.2), supports and allows the *Decision Managers* to classify stakeholders (and users) into four possible categories:

1. **High numerosity, high expertise:** This is the most important category. This class of users and stakeholders is essential for large-scale data acquisition. Indeed, this category is able to provide/generate a large amount of reliable data.
2. **High numerosity, low expertise:** This category of users and stakeholders does not have the expertise to provide totally reliable data. On the other hand, these groups become important for acquiring a large amount of data in absence of users classified in the first category—"**High numerosity, high expertise**".
3. **Low numerosity, high expertise:** Even in this case, this category of users and stakeholders became important in absence of the first category. Indeed, this class is able to provide reliable data even if the number of users does not allow a large-scale acquisition.
4. **Low numerosity, low expertise:** This category classifies all the users and stakeholders who are unable to provide neither reliable data nor in large quantities.

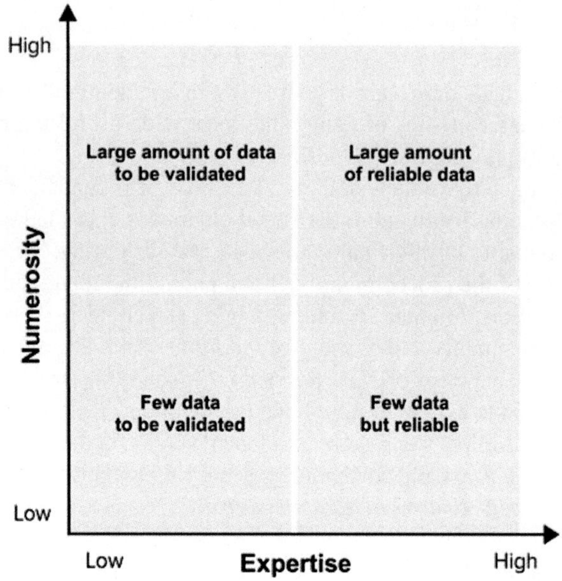

Fig. 4.2 *Numerosity/Expertise grid* to support the users and stakeholders' classification

To this aim, this category must be involved in data acquisition only in cases when there is a lack of other types of stakeholders.

Note that in the **First Case (*"User Reporting supported decision making"*)** *user reporting* is developed to reach a large number of DMs to perform an MCA. Consequently, the *users* identified among the stakeholders are the DMs of the decision problem.

4.3 Selection of Technological Tools to Support the Acquisition of Data (e.g. Smart Devices)

A fundamental support of *User Reporting* is offered using modern technologies. Technological tools can be essential to improve the data acquisition process. Indeed, an effective selection of technological supports allows acquiring a large quantity of data by saving time and involving a large set of users.

The technologies that can support the data acquisition are numerous and the continuous technological evolution offers new possibilities every year. The rationality of the choice can exploit the results of the *Numerosity/Expertise grid*. In particular, the technology can improve the *numerosity* (quantity of users which can report data) or improve the *expertise* (quality of acquired information) of the user during the data acquisition.

To improve the "*Numerosity*", technologies such as *smart devices* can be used together with specific application software to reach a large number of users and increase the quantity of the collected data (participatory sensing).

To improve the "*Expertise*" and consequently the quality of the acquired data, the *smart devices* can provide useful information and a guided data acquisition process to the user. Beyond *smart devices,* the quality of data can be improved with quantitative information acquired by smart sensors networks (*High-resolution cameras, thermology, radiography, ultrasonic, acoustics, elastic*).

In what follows we give two examples later discussed in Chap. 5:

(i) An application (*APP) for smart devices* can be developed to reach a large number of users or to increase the quality of the collected data by providing useful information and a structured data acquisition process. A large variety of APPs can be developed and customized according to the problem to be investigated. The users' reports can be correlated and enriched with questionnaires, photo, video, audio, accelerometer registration, depending on the power and availability of the devices used (example showed in Chap. 5, Sect. 5.1).

(ii) *Augmented Reality* can be used to support the quality of the collected data by proposing the AR-AHP to increase the level of user expertise (by providing a large amount of information during the decision process). Beyond the use of AR-AHP (widely discussed in Chap. 3), *Augmented Reality* can be employed to support the user data acquisition in many fields including emergency procedures, interior design, design on different levels (building, urban district, regional area). In addition, in the case of some extreme event (e.g. seismic or pandemic events), the AR can support its investigation by simulating an on-site survey (Sangiorgio et al., 2021).

On the other hand, other technologies can be used to complete the information of *User Reporting.* These tools do not give direct support to the user but provide complementary information to the *User Reporting,* and they are useful for the final goal of the analysis. These technologies are not analyzed in detail in this book but the most used in building construction management are reported. Here are some examples:

(iii) *Modern high-resolution cameras and image recognition* can investigate superficial damages or construction failure;

(iv) *Modern photogrammetry* allows obtaining reliable information about physical objects and the environment through the process of recording, measuring and interpreting photographic images and patterns of electromagnetic radiant imagery and other phenomena.

(v) *Thermology, radiography, ultrasonic, acoustics, elastic waves* can give useful information about the non-visible part of the structures or materials;

(vi) *Dynamic parameters sensors* can identify the dynamic and static behaviour of civil structures;

(vii) *Strain measurements* typically is the principal feature considered for structural monitoring. The knowledge of strains helps to understand the structural behaviour and the possible causes of damage, such as static or dynamic loads, temperature, etc.

4.4 Creation of Questionnaires, Interviews and Choice Experiments

The creation of the questionnaires is strictly dependent on the *User Reporting's* purposes in synergy with the MCA (Remember the two options defined at the beginning of Chap. 4: *User Reporting supported decision making* to include numerous DMs; or *User Reporting supported large-scale analysis* where an MCA is previously developed).

It is possible to define a general recommendation and two specific guidelines related to the *User Reporting*'s purposes:

(i) In general, it is recommended that a specific questionnaire or interview be developed for every cluster of users in order to calibrate the questions on the basis of the skills and background of the users.
(ii) In the **First Case** (**"*User Reporting supported decision making*"**) it is essential that the users extract the necessary information about their perception in order to carry out the MCA. This approach can be employed in every typology of MCA by defining questionnaires able to provide all the necessary information to perform the weighting in the MCA. In this book, in order to provide an example, the case of the AHP is considered. The questionnaires are defined to fill the matrix of comparisons by using the user's answers in order to evaluate the weights. In the example of the AHP, the questionnaires can be created to allow the user the ability to provide the necessary information to fill the matrix of comparisons.

Let us suppose that the generic matrix of comparison G (Fig. 4.3) need to be evaluated for a large set of users.

The values $a_{1,2}$, $a_{1,3}$, $a_{2,3}$ need to be assigned by using the verbal scales and the associated to numerical values (Equal importance 1, Moderate importance of one over another 3, Strong importance 5, Very strong importance 7, Extreme importance 9).

In this case, the questionnaires can be composed of 3 questions as it is explained in the following.

The question to obtain $a_{1,2}$ "Which is the most important *criterion* between A and B?".

- Equal importance
- A *Moderately* more important than B
- A *Strongly* more important than B
- A *Very Strongly* more important than B
- A *Extremely* more important than B
- B *Moderately* more important than A
- B *Strongly* more important than A
- B *Very Strongly* more important than A
- B *Extremely* more important than A

Matrix G	A	B	C
A	1	$a_{1,2}$	$a_{1,3}$
B	$1/a_{1,2}$	1	$a_{2,3}$
C	$1/a_{1,3}$	$1/a_{2,3}$	1

Fig. 4.3 Generic matrix of comparison G

The question to obtain $a_{1,3}$ "Which is the most important *criterion* between A and C?".

- Equal importance
- A *Moderately* more important than C
- A *Strongly* more important than C
- A *Very Strongly* more important than C
- A *Extremely* more important than C
- C *Moderately* more important than A
- C *Strongly* more important than A
- C *Very Strongly* more important than A
- C *Extremely* more important than A

The question to obtain $a_{2,3}$ "Which is the most important *criterion* between B and C?".

- Equal importance
- B *Strongly* more important than C
- A *Strongly* more important than C
- B *Very Strongly* more important than C
- B *Extremely* more important than C
- C *Moderately* more important than B
- C *Strongly* more important than B
- C *Very Strongly* more important than B
- C *Extremely* more important than B

Note that even numbers (2, 4, 6, 8) could be used as compromises by using the following additional verbal judgments (Weak or Slight, Moderate Plus, Strong Plus, Very very strong) respectively (Afshari et al., 2010).

In the case of the use of even numbers as compromises, the fundamental scale of absolute numbers is improved as reported below:

$a_{i,j} = 1$ Equal importance;
$a_{i,j} = 2$ *Weak or Slight importance of one over another*;
$a_{i,j} = 3$ Moderate importance;
$a_{i,j} = 4$ *Moderate Plus importance*;
$a_{i,j} = 5$ Strong importance;
$a_{i,j} = 6$ *Strong Plus importance*;
$a_{i,j} = 7$ Very strong importance;
$a_{i,j} = 8$ *Very very strong importance*;
$a_{i,j} = 9$ Extreme importance.

Alternatively, it is possible to directly ask the numerical value of the comparison as reported as follows.

Which of the two *criteria* is more important? And how much?

Answer to obtain $a_{1,2}$:

- A = B
- A = 3 B
- A = 5 B
- A = 7 B
- A = 9 B
- B = 3 A
- B = 5 A
- B = 7 A
- B = 9 A

Answer to obtain $a_{1,3}$:

- A = C
- A = 3 C
- A = 5 C
- A = 7 C
- A = 9 C
- C = 3 A
- C = 5 A
- C = 7 A
- C = 9 A

Answer to obtain $a_{2,3}$:

- B = C
- B = 3 C
- B = 5 C
- B = 7 C
- B = 9 C
- C = 3 B
- C = 5 B
- C = 7 B
- C = 9 B

Note that also in this case, even numbers (2, 4, 6, 8) can be used as compromises by defining additional possible answers (e.g. A = 2B, A = 4B, A = 6B and so on).

(iii) In the **Second Case (“*User Reporting supported large-scale analysis*”)** the questionnaires, interviews or choice experiments are not directly related to the realization of the weighting (e.g. the generation of the matrix of comparison to involve many users in the AHP). On the contrary, the questionnaires are used to acquire other types of information that can be useful for applying the results of the MCA on a large scale. In this case, the acquired data must satisfy the data required to apply the results of the MCA (specific *KPIs* or *Cr*). In this case, the acquired information should be sufficient for the calculation of the specific KPIs or condition ratings. An example related to this second purpose (including condition ratings) is showed in Chap. 5 regarding the building degradation analysis and maintenance on the coast of Valencia, Spain.

4.5 Definition of the Flowchart Describing the Acquisition Process

After the questionnaires are expressed, the flowchart describing the *User Reporting* procedure can be defined. An accurate flowchart offers the possibility of clarifying the steps of *Effective Data acquisition*; *Data processing and validation* and *Data analysis* (steps 6, 7 and 8 respectively) of the procedure for the decision problem considered.

In order to better specify the connection and the flow of data among technological tools, *users* and stakeholders, a suitable activity diagram (flowchart) can be designed according to the Unified Modelling Language (UML) framework (Miles & Hamilton, 2006). The UML activity diagram reports in the columns the cluster of *users*, stakeholders and technologies that perform the activities. In turn, the activities are listed in the corresponding columns. Indeed, the flowchart describes and emphasizes the possibility of bringing together huge amount of available data acquired by exploiting "**high numerosity** *users*" and the useful experience of "**high expertise** *users*".

The diagram can include Start node, Action states, Control Flows, Decision nodes, Forks, Joints and End state according to Fig. 4.4.

> **Note** that it is important to consider a validation action in a path of the acquired data in the eventual reporting of low expertise clusters of users. Indeed, any low expertise clusters of users can report data that can be validated by other

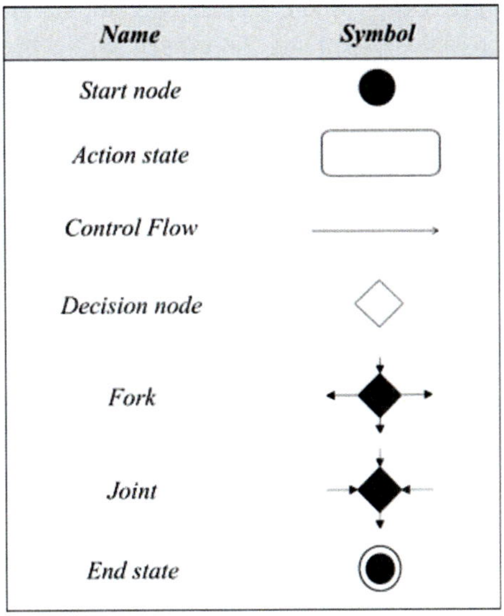

Fig. 4.4 Start node, action states, control flows, decision nodes, forks, joints and end state according to the (UML) framework

clusters of users or stakeholders with high expertise (and consequently able to confirm or modify the data on the basis of their knowledge).

4.6 Effective Data Acquisition

Once the flowchart is defined, the operative data acquisition can start. In this phase, it this important to include in the acquisition process specific registry data in order to correctly storage the information on the basis of different criteria. In particular, the registry data acquisition could include useful information about the users in the datasets (personal data and information about the expertise level). In addition, in the field of the construction sector, registry data can be very helpful to consider some additional information on the buildings investigated in the analysis. To provide an example, registry data can identify useful information to describe the geometric, morphological and constructive features of the building, including design prescriptions, reference building codes and possible constructive details. Such data, together with the questionnaires and the additional information obtained by the technological tools, complete the "User Reporting" supported data acquisition process.

4.7 Data Processing

The *Data Processing Phase* is based on suitable automatic classification of acquired information which is processed and classified into a database. For this phase, it is possible only to suggest generic guidelines since Data Processing strictly depends on the problem considered.

In the User reporting applied to the **First Case,** the information collected is stored and classified to support the MCA involving numerous users (DMs). The information is prepared to be used to extract weights. In the example of the AHP, the numerical values associated with the qualitative judgments expressed by the users are prepared to be employed in the matrices of comparisons.

When the *User Reporting* is applied to the **Second Case,** data are classified in order to be useful in a validation process or to obtain a large test site along with the results of an MCA. In this case, specific KPIs, condition ratings or risk indexes are used.

4.8 Data Analysis

Data Analysis Phase exploits the formulation and procedure of the MCA specific Key Performance Indicators (KPIs), Condition Ratings (Cr) or risk indexes.

In the **First Case** the data analysis consists of the weighting evaluation and aggregation of the results of the large-scale MCA. Since the AHP is used, the data analysis follows the steps of Saaty's classical approach: *The local weights evaluation, The local consistency test, The global weights evaluation and, The sensitivity analysis.*

Note that the *User Reporting* supported MCA is able to reach a statistically significant number of DMs in order to validate the final decision and achieve a final ranking shared and widely accepted.

In the **Second Case** the *Data Analysis Phase* exploits any Key Performance Indicators (KPIs) or Condition Ratings (*Cr*) to perform large-scale analysis, test site in order to achieve the final goal of the *User Reporting*. An **example of exceptional academic value** is proposed in the next chapter for the goal of t building maintenance and diagnostics on the coast of Valencia, Spain.

4.9 Example: The Simplified Risk Evaluation in the Field of Building Construction Management

In this subsection, a simple explanation of the proposed approach from a practical point of view is proposed to better clarify the use of the *User Reporting*. The proposed example is the same used to show the application of the AHP (Sect. 2.6, Chap. 1). Remember that this example concerns the simplified risk evaluation in the field of building construction (in the case of the heavy expulsion of concrete cover and rebars oxidation in the residential building "*Risk of falling objects*"). The three ***alternatives*** (Fig. 2.8, Sect. 2.7.1) to be ranked are:

(A) *Two men* with *sombrero hat* and block falling from *3 m*,
(B) *One man* with *helmet* and block falling from *6 m*,
(C) *Three children* without head cover (*nothing*) and block falling from *1 m*.

In particular, the following subsections show the eight steps to set and achieve the *User Reporting* applied for the **First Case**.

Note that beyond this simplified example, two real applications of the *User Reporting* in the field of construction are shown in the next Chaps. 6and7 for both the cases.

The **preliminary step** of the *User Reporting* is to identify the goal, the necessary information to be acquired and any MCA to be included in the approach. In this example, the purpose is to perform a large MCA by considering a large set of expert users in the field of construction as DMs to evaluate the risk of injury for the building users due to the detachment of the building concrete cover (*Risk of falling objects*). The necessary information to be acquired is the expert judgment of the users. This information is used to extract weights by using the AHP that is the MCA included in the *User Reporting*.

4.9.1 Example: Identification of the Stakeholders

In this simplified example, the stakeholders identified to consist of 4 classes.

High power, highly interested people are the *Building Managers* and maintainers, i.e. who needs to carry out the analysis to establish which one of the three *alternatives* (A, B and C) is riskiest.
High power, less interested people are identified in the interviewed as *Building Engineers* with experience in the field of maintenance and diagnostics of reinforced concrete buildings (in this case these are also the DMs).
Low power, highly interested people are identified in the *Buildings Users* that may suffer the injuries.
Low power, less interested people are represented by *Local Residents,* but who are not users of the buildings investigated.

4.9.2 Example: Definition of Users

Starting from the defined stakeholders, potential users are defined and classified according to the *Numerosity/Expertise grid* as follows.

High numerosity, high expertise stakeholders are represented by the *Building Engineers.*
High numerosity, low expertise cluster can be represented by the *Buildings Users* who are not able to provide reliable data (these are not identified as the users of the *User Reporting*).
Low numerosity, high expertise stakeholders are identified in the *Building Managers.*
Low numerosity, low expertise cluster can be represented by other *Local Residents* who may not be able to provide useful information.

The left grid of Fig. 4.5 shows the Power/interest grid used to classify stakeholders and the right grid shows the Numerosity/Expertise grid used to group users.

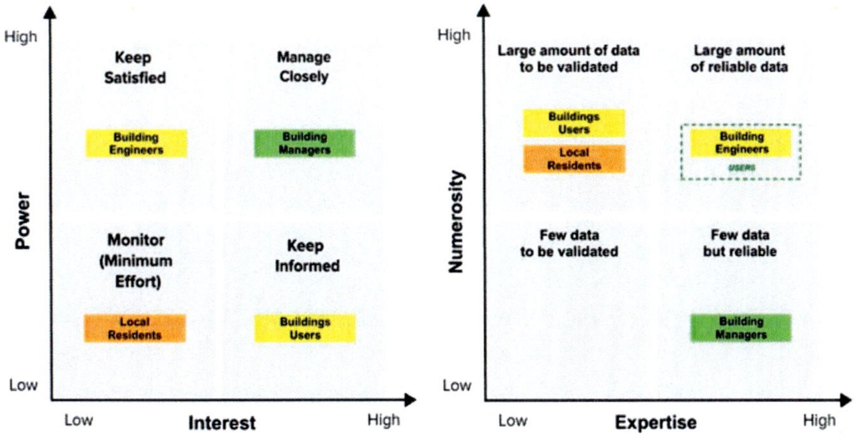

Fig. 4.5 (Left) Power/interest grid used to classify stakeholders, (right) numerosity/expertise used to cluster users

In this example, the cluster of *Building Engineers* has good characteristics both in terms of numerosity and expertise. To this aim the *users* of the *User Reporting* are identified in this cluster of stakeholders.

4.9.3 Example: Selection of Technological Tools to Support Data Acquisition (e.g. Smart Devices)

Among the available technological tools to support the *User Reporting*, a suitable *Web Based Platform* can be used to support the acquisition of expert judgments from the stakeholders and implement the AHP model. *Web Based Platform* in this case, can be a web based Docs Editors suite such as the Google Forms software offered by Google.

By exploiting the *Numerosity/Expertise grid* the operators (users) of the *Web Based Platform* are identified among the *Building Engineers*. In this way, a large number of *Building Engineers* can employ the *Web Based Platform* to execute the AHP. Consequently, the *Building Managers* can exploit the results.

The other classes of users and stakeholders are not needed at this time and consequently are not involved in the *User Reporting* data acquisition.

Matrix A_1: (head cover)	Sombrero	Helmet	Nothing
Sombrero	1	$a_{1,2}$	$a_{1,3}$
Helmet	$a_{2,1}$	1	$a_{2,3}$
Nothing	$a_{3,1}$	$a_{3,2}$	1

Fig. 4.6 *Judgment Matrix* A_1 related to the *criterion Vulnerability* (head cover)

4.9.4 Example: Creation of Questionnaires

The questionnaires are developed to extract the necessary information from the *Building Engineers,* exploiting their expertise, to carry out the AHP. The questionnaires are set to be answered by users by using the *Web Based Platform.* The information acquired by the questionnaires needs to be used to fill the *Judgment Matrices* of the AHP. In the proposed example, the *matrix* A_1 (Fig. 4.6) is the first matrix to be filled, related to the criterion *Vulnerability* (head cover).

In this case, the questionnaires consist of the following 3 questions.

Question to obtain $a_{1,2}$: which is the most "vulnerable" between *Sombrero* (S) and *Helmet* (H)?

- Equal vulnerability
- S *Moderately* more vulnerable than H
- S *Strongly* more vulnerable than H
- S *Very Strongly* more vulnerable than H
- S *Extremely* more vulnerable than H
- H *Moderately* more vulnerable than S
- H *Strongly* more vulnerable than S
- H *Very Strongly* more vulnerable than S
- H *Extremely* more vulnerable than S

Question to obtain $a_{1,3}$: which is the most "vulnerable" between *Sombrero* (S) and *Nothing* (N)?

- Equal vulnerability
- S *Moderately* more vulnerable than N
- S *Strongly* more vulnerable than N
- S *Very Strongly* more vulnerable than N
- S *Extremely* more vulnerable than N
- N *Moderately* more vulnerable than S
- N *Strongly* more vulnerable than S
- N *Very Strongly* more vulnerable than S
- N *Extremely* more vulnerable than S

Question to obtain $a_{2,3}$: which is the most "vulnerable" between *Helmet* (H) and *Nothing* (N)?

- Equal vulnerability
- H *Moderately* more vulnerable than N
- H *Strongly* more vulnerable than N
- H *Very Strongly* more vulnerable than N
- H *Extremely* more vulnerable than N
- N *Moderately* more vulnerable than H
- N *Strongly* more vulnerable than H
- N *Very Strongly* more vulnerable than H
- N *Extremely* more vulnerable than H

Alternatively, it is possible to ask directly for the numerical value of the comparison as follows.

Answer to obtain $a_{1,2}$:

- S = H
- S = 3 H
- S = 5 H
- S = 7 H
- S = 9 H
- H = 3 S
- H = 5 S
- H = 7 S
- H = 9 S

Answer to obtain $a_{1,3}$:

- S = N
- S = 3 N
- S = 5 N
- S = 7 N
- S = 9 N
- N = 3 S
- N = 5 S
- N = 7 S
- N = 9 S

Answer to obtain $a_{2,3}$:

- H = N
- H = 3 N
- H = 5 N
- H = 7 N
- H = 9 N
- N = 3 H
- N = 5 H
- N = 7 H
- N = 9 H

4.9.5 Example: Definition of the Flowchart Describing the Acquisition Process

At this point, the stakeholders, users, technologies and questionnaires are defined, consequently, the UML activity diagram of the procedure can be sketched to clarify the whole process of the *User Reporting* for the proposed example (Fig. 4.7).

First, the *Building Engineer* provides registry data (e.g. their user id) in the *Web Based Platform*. Consequently, the *Web Based Platform* registers the user and provides the questionnaire with all the questions required to complete the AHP. Once the *Building Engineer* answers all the questions, the "report" is stored in the *Web Based Platform* that prepares data for the AHP evaluation. At this point, the *Web Based Platform* performs the weighting according to the classical procedure of the AHP. If the result does not satisfy the consistency test (see Sect. 2.3), the *Building Engineer* is asked to perform the questionnaires until the consistency is reached. When the consistency is satisfied, the weights of the *Building Engineer* are stored. After all the *Building Engineers* provides consistent weights evaluation, the *Web Based Platform* can aggregate the outcomes to obtain the final results (the ranking of the riskiest situation between the three ***alternatives*** A, B and C). Note that the UML activity diagram describes the steps of *Effective Data acquisition*; *Data processing*

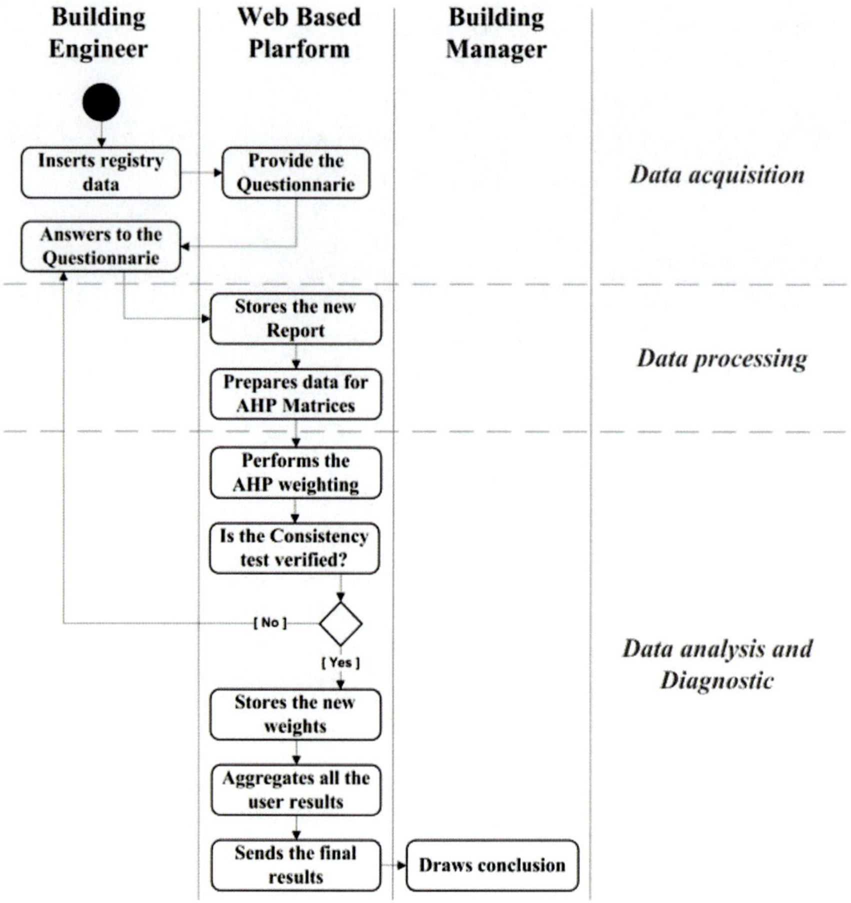

Fig. 4.7 UML activity diagram of the *User Reporting* applied to the example of "Risk of falling objects"

and validation, and *Data analysis* (steps 6, 7 and 8, respectively) which are detailed in the following subsections.

4.9.6 Example: Effective Data Acquisition

This subsection shows the example of the *Building Engineer* that answers the part of the questionnaire related to the Judgment Matrix A_1 (criterion Vulnerability of head cover). Such answers are reported in Fig. 4.8.

Consequently, the values associated with matrix A_1 (Fig. 4.8) are obtained and reported as follows:

$a_{i,j}$	*Criterion*	Which of the two *criteria* (head covers) is more vulnerable? And how much?									*Criterion*
$a_{1,2}$	*Sombrero* (S)	☐ S=9H	☐ S=7H	☐ S=5H	☑ S=3H	☐ S=H	☐ H=3S	☐ H=5S	☐ H=7S	☐ H=9S	*Helmet* (H)
$a_{1,3}$	*Sombrero* (S)	☐ S=9N	☐ S=7N	☑ S=5N	☐ S=3N	☐ S=N	☐ N=3S	☐ N=5S	☐ N=7S	☐ N=9S	*Nothing* (N)
$a_{2,3}$	*Helmet* (H)	☐ H=9N	☐ H=7N	☐ H=5N	☑ H=3N	☐ H=N	☐ N=3H	☐ N=5H	☐ N=7H	☐ N=9H	*Nothing* (N)

Fig. 4.8 Answers to the questions to evaluate the criterion Vulnerability of head cover

$$a_{1,2} = 3$$
$$a_{1,3} = 5$$
$$a_{2,3} = 3$$

4.9.7 Example: Data Processing

To provide the same example, once the data are acquired and stored, the information can be organized in a matrix. Indeed, the matrix of comparison A_1 can be obtained by including the questionnaire answers (converted into numerical judgments) in the matrix in the correct cell as shown in Fig. 4.9.

4.9.8 Example: Data Analysis

In the last step of the *User Reporting* procedure the local weights can be evaluated, the consistency test can be performed and the global weights evaluated according to the classical AHP.

Continuing with the same example, Fig. 4.10 shows the results of the matrix of comparisons A_1.

After all the *Building Engineers* filled the questionnaire and all the weights are evaluated, it is possible to obtain the final results of the analysis. Consequently, the

Matrix A_1: (head cover)	Sombrero	Helmet	Nothing
Sombrero	1	*3*	*5*
Helmet	*1/3*	1	*3*
Nothing	*1/5*	*1/3*	1

Fig. 4.9 Weights extracted from the questionnaire to fill the *Judgment Matrix A_1* related to the criterion *Vulnerability* (head cover)

Matrix A_1: (head cover)	Sombrero	Helmet	Nothing	CR	Weights
Sombrero	1	*3*	*5*		*0.63*
Helmet	*1/3*	1	*3*	*0.04 (<0.1) Consistent*	*0.26*
Nothing	*1/5*	*1/3*	1		*0.11*

Fig. 4.10 Results of the Judgment Matrix A_1

Building Managers can identify the riskiest situation among the three ***alternatives*** A, B and C by aggregating the global weight of every user (DM).

Note that the results of an AHP aggregating a large number of experts is much more reliable and it is necessary to demonstrate that the final ranking is widely accepted. Indeed, statistical analyses can also be performed to demonstrate the convergence of all DMs to the final result (Caporale et al., 2020). Moreover, there is a vast number of different methods and possibilities to aggregate the individual DM preferences to a group consensus. An exhaustive review on various aggregation possibilities is offered by Ossadnik et al. (2016).

References

Afshari, A., Mojahed, M., & Yusuff, R. M. (2010). Simple additive weighting approach to personnel selection problem. *International Journal of Innovation, Management and Technology, 1*(5), 511.

Caporale, D., Sangiorgio, V., Amodio, A., & De Lucia, C. (2020). Multi-criteria and focus group analysis for social acceptance of wind energy. *Energy Policy*. 140111387. https://doi.org/10.1016/j.enpol.2020.111387.

Miles, R., & Hamilton, K. (2006). *Learning UML 2.0: a pragmatic introduction to UML*. O'Reilly Media, Inc.

Ossadnik, W., Schinke, S., & Kaspar, R. H. (2016). Group aggregation techniques for analytic hierarchy process and analytic network process: a comparative analysis. *Group Decision and Negotiation, 25*(2), 421–457. https://doi.org/10.1007/s10726-015-9448-4

Sangiorgio, V., Iacobellis, G., Adam, J. M., Uva, G., & Fatiguso, F. (2018a). User-Reporting based decision support system for reinforced concrete building monitoring. In *2018 IEEE international conference on systems, man, and cybernetics (SMC)* (pp. 2620–2625). IEEE.

Sangiorgio, V., Uva, G., & Fatiguso, F. (2018b). User reporting–based semeiotic assessment of existing building stock at the regional scale. *Journal of Performance of Constructed Facilities, 32*(6), 04018079.

Sangiorgio, V., Pantoja, J. C., Varum, H., Uva, G., & Fatiguso, F. (2019a). Structural degradation assessment of RC buildings: Calibration and comparison of semeiotic-based methodology for decision support system. *Journal of Performance of Constructed Facilities, 33*(2), 04018109.

Sangiorgio, V., Uva, G., Fatiguso, F., & Adam, J. M. (2019b). A new index to evaluate exposure and potential damage to RC building structures in coastal areas. *Engineering Failure Analysis, 100*, 439–455.

Sangiorgio, V., Martiradonna, S., Fatiguso, F., & Lombillo, I. (2021). Augmented reality based-decision making (AR-DM) to support multi-criteria analysis in constructions. *Automation in Construction, 124*, 103567.

Sangiorgio, V., Uva, G., & Aiello, M. A. (2020). A multi-criteria-based procedure for the robust definition of algorithms aimed at fast seismic risk assessment of existing RC buildings. In *Structures* (Vol. 24, pp. 766–782). Elsevier.

Chapter 5
AR-AHP to Support the Building Retrofitting: Selection of the Best Precast Concrete Panel Cladding

Abstract *Augmented Reality* in combination with the *Analytic Hierarchy Process* (AR-AHP) is a powerful tool to support researchers, practitioners, engineers and architects in different areas of the building construction sector. One of the most effective applications of the AR-AHP regards building retrofitting, were different parameters (*qualitative* and *quantitative*, *technical* and *non-technical*) are involved. To provide an example, the aesthetics impact is a *qualitative* and *non-technical* criterion that can have a strong influence on both the customer purchase and the local resident's acceptance. On the other hand, *quantitative* and *technical* information such as the structural or thermal characteristics and costs are fundamental for building performance. This chapter proposes a practical application of the AR-AHP in the field of building retrofitting. In particular, the AR-AHP is developed to face the complex problem of the best cladding selection (applied on an experimental Precast Concrete Panels-PCPs) to be used in building retrofitting. In this context, the AR-AHP is particularly effective since the decision makers can be supported by 3D models (displayed in AR) and a simple weights evaluation procedure in order to evaluate the multiple aspects of the PCPs and achieve a decision by considering multiple criteria.

Keywords Precast concrete panel · Case study · Architectural engineering · Building retrofitting · Augmented reality · Analytic hierarchy process · Technological innovation · Simos-Roy-Figueira method · Building 3D modelling

This chapter proposes a practical application of the AR-AHP (theory explained in Chap. 3) in the field of building retrofitting. In this field of application, the AR-AHP is a powerful tool to support researchers, practitioners, engineers and architects. In the proposed application, the goal of the MCA regards the identification of the best solution among a set of *alternatives* (*MCA Goal B* defined in Sect. 1.4). In particular, the AR-AHP is developed to face the complex problem of the best cladding selection (applied on an experimental Precast Concrete Panels-PCPs) to be used in a building retrofitting (Martiradonna et al., 2019, 2020a, b). In this problem, a fundamental choice regards the cover material selection. Indeed, the cover can influence the characteristics of the panel in terms of aesthetical effect, time and needs of production, installation, technical properties and economic sustainability.

V. Sangiorgio et al., *New Approaches for Multi-Criteria Analysis in Building Constructions*, https://doi.org/10.1007/978-3-030-83875-1_5

The PCPs system is a novel technological solution not yet well known among the technicians of the building sector (Sangiorgio et al., 2021). Consequently, several involved *criteria* and *alternatives* are difficult to be compared by using a classical AHP because the DM will simultaneously consider a great amount of information about the novel PCPs system to make a consistent decision.

In this context, the AR-AHP is particularly effective since the DM can be supported by 3D models displayed in AR and a simple weights evaluation procedure to understand and evaluate the multiple aspects of the PCPs and achieve a decision by considering multiple criteria.

5.1 The Building to be Renovated with the PCPs System

The construction to be renovated is a Reinforced Concrete (RC) building of an economic-popular residential complex located in Trani, a city few kilometres far away from Bari, Puglia (Italy). It was built between 1958 and 1963 and consists of a basement and two floors, four apartments and a regular footprint area of 258 m^2. It has two principal facades with lodges and balconies, a lateral blind facade and the last one is shared with another building. The external wall thickness is 25 cm in the lodges and 40 cm in the rest of the building with a thermal transmittance value (U-Value) of 1.03 W/m^2K. The original envelope is realized in stone tiles in the basement and plastered in the upper part with no insulating layers included. The specific multi-criteria problem regards the section of the best PCP to be used in the building retrofitting among a set of six alternatives: *Exposed Concrete*, *Wood Coated*, *Brick Coated*, *Ceramic Coated*, *Stone Coated* and *Plastered*. Figure 5.1 shows a 3D model of the building to be renovated after the possible application of the *Plastered* PCP.

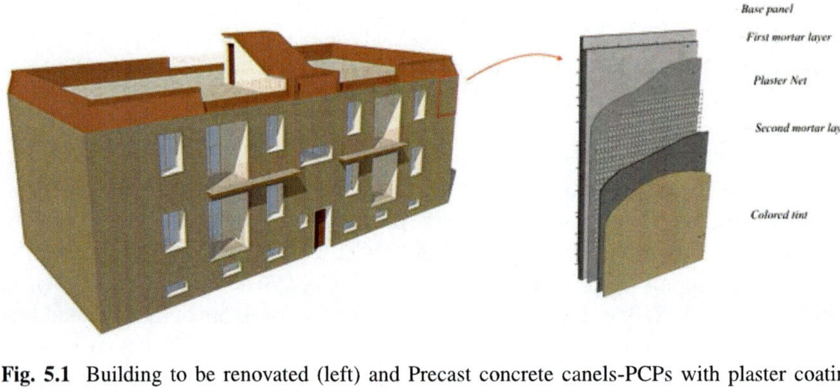

Fig. 5.1 Building to be renovated (left) and Precast concrete canels-PCPs with plaster coating (right)

Moreover, the DMs need to contemplate the following criteria in the decision problem: *Aesthetic Impact*, *Required construction vehicles*, *Realization Complexity*, *Thermal Transmittance*, *Thermal Inertia* and *Expected Cost*.

The DMs are represented by a group of practitioners (designers) operating in the construction field. Since the PCP is a novel and experimental technology, the decision is complicated by the absence of previous applications both in practical and scientific scenarios. The AR-AHP can be effective to better understand the technology and to visualize in augmented reality what the building would look like, directly on site, after applying the panels.

5.2 AR-AHP Step 1, *Structuring the Problem* for the PCP Selection

The AR-AHP step 1 consists in the *Structure of the Problem* to determine an effective choice regarding the best selection of the PCP cladding for the building envelope.

In particular, the goal is defined as the *Precast Concrete Panel Cladding Selection*. To this aim, as already mentioned in the previous section, six *criteria i* (with $i = 1, \ldots, 6$) are defined: (1) *Aesthetic Impact*; (2) *Required construction vehicles*; (3) *Realization Complexity*; (4) *Thermal Transmittance*; (5) *Thermal Inertia;* (6)*Expected Cost*.

Moreover, the *criteria* are grouped into four *macro-criteria*: *aesthetics*, *production and executive needs*, *thermal behaviour*, and *costs*.

Note that the grouping in *macro-criteria* have no influence in the AR-AHP procedure and weighting but it is proposed only to better categorize and display the criteria in the hierarchical flowchart (*Structuring the problem*).

In addition, a set of six different smart precast envelope solutions are proposed as *alternatives j* (with $j = 1, \ldots, 6$) of the decision problem: (1) *Exposed Concrete*; (2) *Wood Coated*; (3) *Brick Coated*; (4) *Ceramic Coated*; (5) *Stone Coated*; (6) *Plastered*. The six *criteria* and the six possible *alternatives* are structured in a hierarchical flowchart that is shown in Fig. 5.2. The flowchart displays how each *alternative* is related and then connected with all the defined *criteria*.

All the defined *criteria* and *alternatives* are justified according to the needs and demands of the building to be renovated and discussed as follows.

The first *macro-criterion* and *criterion* regard the aesthetic aspects of the panel in relation to the case study.

(1) *Aesthetic Impact* ($i = 1$) has a strong influence both on the customer purchase and the local resident's acceptance. Indeed, Pittau et al. (2017) argue that a high aesthetic impact may return dignity to the building, integrate tenants with the rest of the city and make the building gain value. Indeed, typically an aesthetic

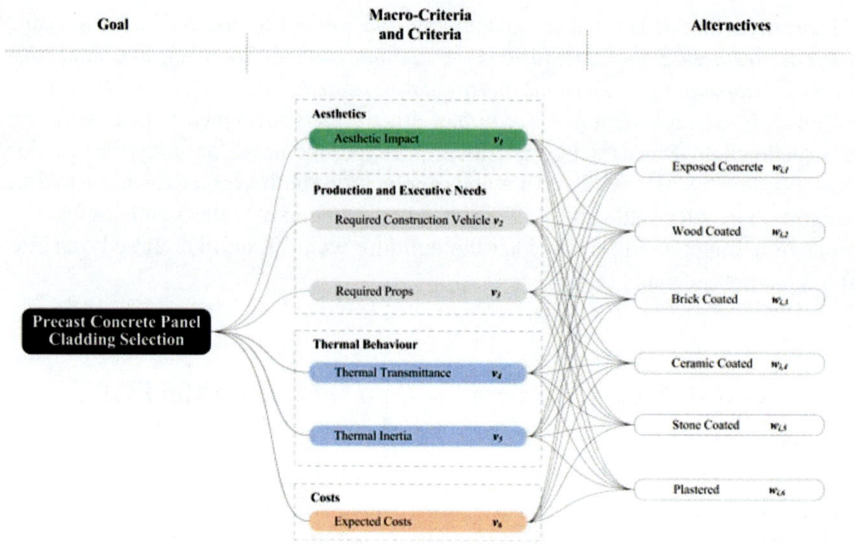

Fig. 5.2 *Structuring the problem*: *criteria* and *alternatives* to determine the best smart precast envelop

similarity to the rest of the urban context is preferable and a low aesthetic impact may have some negative effects, i.e. marginalization and social discrimination.

The second *macro-criterion* takes the *Production and Executive Needs* into account to evaluate the PCP cladding. It considers two qualitative *criteria,* one related to the vehicle required to transport the panels from the factory to the construction site and another to evaluate the realization complexity required during the prefabrication and installation process of the panels. To this aim, the second and third *criteria* are defined.

(2) *Required construction vehicles* ($i = 2$) is related to the typology of lorry employed to deliver the panels to the construction site to respect the Italian regulatory (D.P.R., 1992). The type of used vehicle and the number of trips depend on the arrangement and the maximum number of panels. In turn, they depend on the size and weight of the panels.

(3) *Realization Complexity* ($i = 3$) is important to consider the different requirements and measures for the correct production and execution of the panels, both in the factory and in the construction place (according to the rules of workers prevention and protection on the work).

The third *macro-criterion* is the *Thermal Behaviour* of the building, which can be assessed through two important quantitative *criteria*:

(4) *Thermal Transmittance* ($i = 4$) is the coefficient of heat transmission between surfaces. The regulation UNI EN ISO 6946 (2018) defines thermal transmittance as "the heat flow through a unit surface subjected to a temperature difference of one degree".

(5) *Thermal Inertia* ($i = 5$) is the capacity of a building component to mitigate the temperature fluctuations in the internal environment due to the variation of thermal loads throughout the day. This characteristic is particularly important for the overall thermal performance of the building, specifically in southern Italy (Boeri & Longo, 2011).

The final *macro-criterion* considers a preliminary evaluation of *Costs*. To this aim, the last quantitative *criterion* is defined as follows:

(6) *Expected Cost* ($i = 6$) is the economic parameter that considers the increase of expenses to realize the PCP depending on the cladding material and its assembling technology.

Supposed that the retrofitting technology of the PCPs system remains unchanged, six PCP *alternatives* maybe defined at the cladding variation.

(1) *Exposed Concrete* ($j = 1$) is the base panel (PCP) of the experimental retrofitting system (1.2 m wide, 2 m long and 0.1 m thick). This is realized by a precast reinforced concrete plate including an internal insulating layer in lightweight mortar and recycled EPS blocks. Its external finish shows the typical grey colour of Portland cement concrete and presents a smooth outer surface.

(2) *Wood Coated* ($j = 2$) is the PCP integrated with the wooden planks used for external surface covering. The planks are designed to be easily mounted in the construction site by a mullion-clip system embedded in the external concrete plate of the panels. This technology permits hiding the junctions among the panels.

(3) *Brick Coated* ($j = 3$) is the base panel integrating ceramic solid bricks during the prefabrication process. Indeed, this panel appears as a brick wall. The bricks, 25 cm long, 12 cm wide and 2.5 cm thick, are arranged in rows along the width and with offset vertical joints. Hence, the panels leave the factory already with the brick coat.

(4) *Ceramic Coated* ($j = 4$) is the standard PCP cladded with ceramic tiles. The cladding is carried out in the factory, before casting the concrete of the plate. They can be arranged according to the architectural design and with various shape tiles. Those considered are 20 × 40 cm in brown shadows. Also, in this case, the PCPs leave the factory integrated with the ceramic cladding.

(5) *Stone Coated* ($j = 5$) is the base panel cladded by "Trani' stone" tiles. This stone type is widely used in Apulian region, in Southern Italy, for its characteristics and whitish aspect. In the factory, according to the designer drawing, the tiles are arranged as the first layer of the PCP and integrated with the slab by means of the concrete casting. The stone coated PCP arrives at the construction site ready to be mounted on the building.

(6) *Plastered* ($j = 6$) is the base panel integrated with a set of layers composed of mortar, a plaster net, a second layer of mortar and the coloured tint. After installing the panels and adding the net in the junctions among the panels, the existing building facade is cladded in situ with the plaster, coloured on the client request.

5.3 AR-AHP Step 2, *AR Setting* and 3D Models Design (PCPs and Building)

Once the problem is defined, the AR-AHP step 2 is applied and the 3D models c_j associated with *alternatives* j (for $j = 1, \ldots, 6$) are realized.

In particular, in this multi-criteria problem, six 3D models are developed to show the final effect of the different PCPs applied to the case study in AR useful to investigate the *Aesthetic Impact* criterion. Moreover, other six 3D models are realized to represent the different PCP cladding variations to support the evaluation of *Realization Complexity*. Figure 5.3 shows the 3D model of the *Stone Coated* PCP in the right and the *Stone Coated* PCP applied to the case study in the left.

It is worth noting that the comparison regarding the remaining *criteria* of *Required construction vehicles, Thermal Transmittance, Thermal Inertia and Costs* can be quantitatively carried out. To this aim, a specific 3D model is not necessary to support these specific comparisons.

To obtain the used 3D models, the open access software "**SketchUp** make 2017" is used. Moreover, various free plug-ins are available, useful for modelling and returning models in an AR environment.

In particular, in the following, three steps to set an AR environment with the software **SketchUp** and **Augment** are discussed. **Augment** is an *Augmented Reality* Software as a Service (SaaS) platform that allows users to visualize their models in 3D in real environment and in real-time through tablets or smartphones.

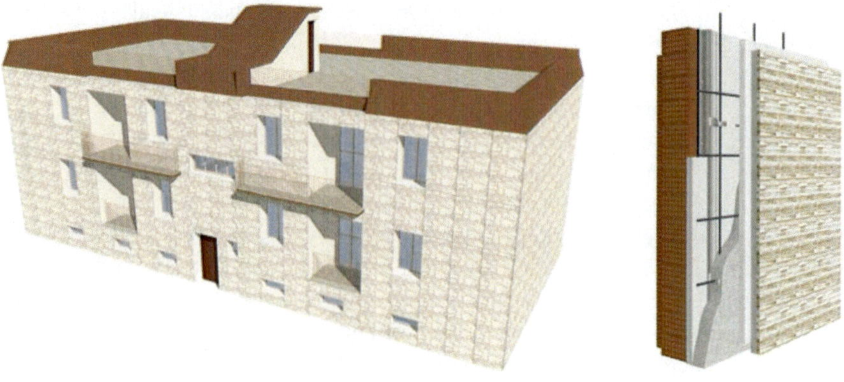

Fig. 5.3 *Stone Coated* PCP: 3D model of the panel and application to the case study

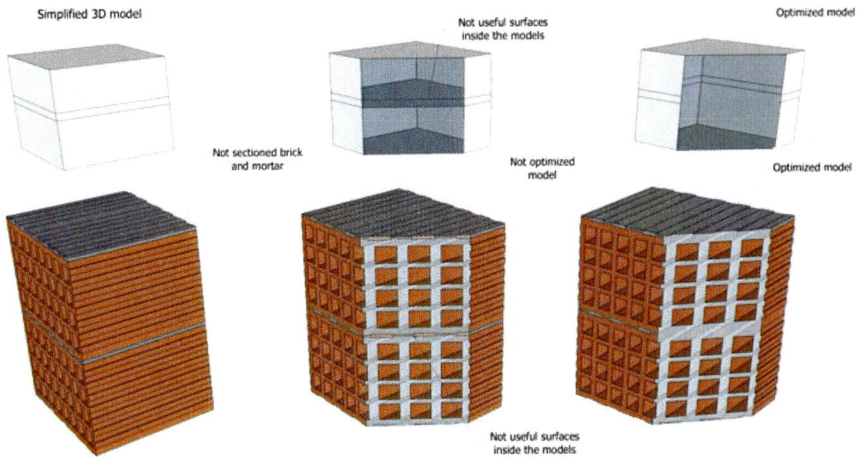

Fig. 5.4 Optimization of the 3D model in **SketchUp** software

(1) The first step is to model the object by a generic 3D modelling software such as **SketchUp**. Note that it is important to avoid creating (not useful) surfaces inside the models that are not visible from the outside to keep small the file size of the 3D model. Heavy files can slow down the visualization in the AR software. In particular, Fig. 5.4 shows how to optimize the model by avoiding not useful surfaces inside the models. In the left part of Fig. 5.4 two bricks with the mortar layer are shown. In the central part of the image, a sectional view shows the internal part of the model with the presence of "not useful surfaces" inside (not optimized model). In addition, in the right part of the image, the optimized model is displayed (without not useful surfaces). The upper part of the image clarifies the optimization process with three additional simplified models. Once all the components are realized, the model is exhibited in the **SketchUp** software as shown in Fig. 5.5.

(2) The second step is to export the 3D model from **SketchUp** and upload it in the AR software such as the **Augment** cloud (it is necessary to create a user profile to upload all your personal models). In general, there are many formats to export a 3D model to be used in an AR environment. An important check regards the model dimension when the file is uploaded in AR software. Indeed, the units and the dimensions may change during the passage from the 3D modelling software to the AR environment. In some cases, the units of a model can be edited in the cloud (or locally) as in the case of **Augment** from the "Augment Manager" platform. In particular, the recommended file extension exportable from **SketchUp** & 3D Warehouse is the **KMZ** or alternatively the STL. It is worth noting that the STL extension is not able to show textures or materials information in the AR. Indeed, by using the STL extension, the 3D model will appear completely blue (or another colour depending on the AR software).

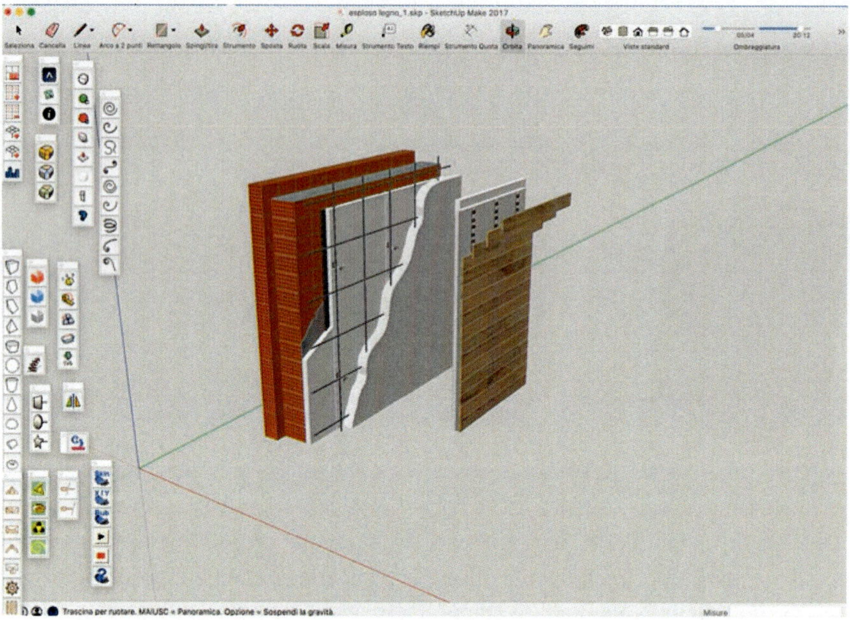

Fig. 5.5 Drawing of the 3D model in **SketchUp** software

In this example the export of a **KMZ** file for modelling with **SketchUp** is proposed (Fig. 5.6).

(3) The third step is the use of the 3D model in the AR environment by exploiting the APP **Augment** for smart devices (Fig. 5.7).

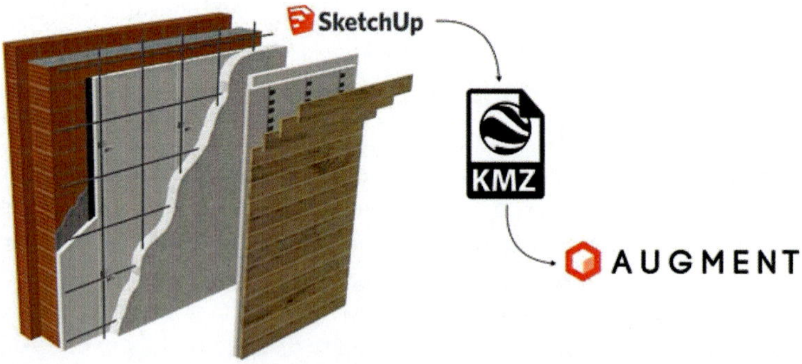

Fig. 5.6 Export of the 3D model from **SketchUp** and upload in the **Augment** cloud

Fig. 5.7 3D model displayed in *Augmented Reality* by using **Augment** software

Note that in general, if the 3D software does not allow exporting in one of the formats DAE, OBJ, STL, KMZ, PLY or Catia, or if there are other issues about a specific format, it is recommended to download the free 3D software **pCon.planner** to convert the model to OBJ or DAE. In particular, software **pCon.planner** allows the following import formats: DWG, DXF, 3DS, GLTF, GLB, PDF, SKP, STL, FBX, IFC, SAT, SAB, FML.

5.4 AR-AHP Step 3, *Local Weights Evaluation*

In AR-AHP step 3 every *criterion* and *alternative* are analyzed to weigh the parameters involved. Let us define the set of *criteria*$N_1 = \{i | i = 1, \ldots, 6\}$ and the set of *alternatives* $N_2 = \{j | j = 1, \ldots, 6\}$. The *criteria* and *alternatives* local weights are defined as follows:

- v_i is the local weight associated with the i^{-th} *criterion* $\forall i \in N_1$;
- $w_{i,j}$ is the local weight associated with the j^{-th} *alternative* related to the $i^{-\text{th}}$ *criterion*, for $\forall i \in N_1, \forall j \in N_2$.

By involving a group of DMs represented by technicians operating in the construction field, the *AR-AHP step 3* allows evaluating local weights: six evaluations are applied to identify the tabulated weights of *alternatives* $w_{i,j}$ and one is developed to evaluate v_i.

Note that if compared with the simplified examples of Chap. 3, the following subsections pay specific attention to the mathematical definition and formalization of the weights.

5.5 AR Phase 1—Local Preliminary Ranking

The DMs start from the evaluation of the *Aesthetic Impact* ($i = 1$) by considering the *alternatives* $j \in N_2$. In *AR phase 1* the DMs order the 3D model associated to *alternatives* j on the basis of their qualitative judgment, starting from the one that has the worst *Aesthetic Impact* to the one that has the best *Aesthetic Impact*. In this phase, the 3D models are ranked from the one with the worst aesthetic impact (the building 3D model obtained with *Wood Coated* PCP) to the one with the best aesthetic impact (the building 3D model obtained with the *Ceramic Coated* PCP).

5.6 AR Phase 2—Local Ranking

Subsequently, in accordance with *AR phase* 2 a set of empty spaces is used to increase the differences between two consecutive *alternatives*. Two empty spaces are included between *Exposed Concrete* and *Plastered* building 3D models and one empty space in included between *Plastered* and *Stone Coated* 3D building 3D models.

Figure 5.8 shows the *Aesthetic Impact* evaluation in AR environment useful to display and compare the final effect of the PCPs applied to the case study. Moreover, in Fig. 5.8 the local ranking obtained by an "on desk" analysis with virtual 3D models in scale is shown.

5.7 AR Phase 3—Local Weights Evaluation

Once the local ranking is obtained for the *Aesthetic Impact* ($i = 1$) criterion, it is possible to extract the relative local weights. The weights $w_{1,j}$ (associated to the *j*-th alternative in relation to the criterion $i = 1$) are determined by using Eq. (3.1) and exploiting the rank assigned and showed in Fig. 5.9.

More in general, local weights $w_{i,j}$ are computed by reworking Eq. (3.1) as follows:

$$w_{i,j} = \frac{r_j}{\sum_{j=1}^{6} r_j}, \forall j \in N_2, \forall i \in N_1. \tag{5.1}$$

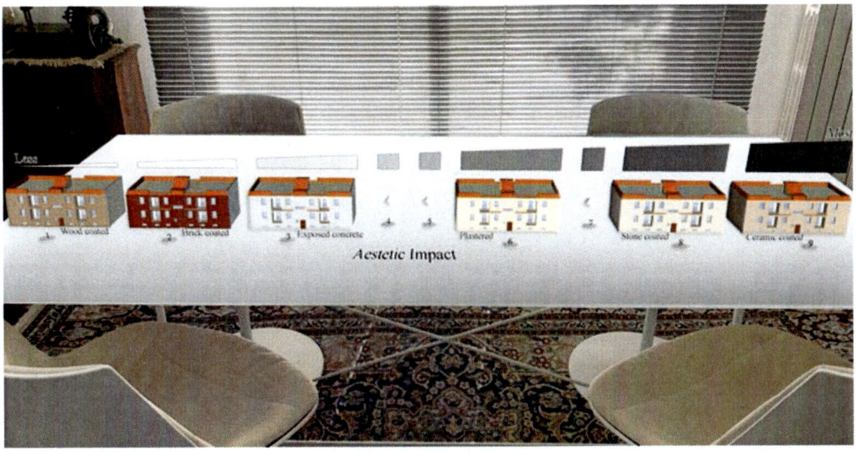

Fig. 5.8 Building 3D models displayed in *Augmented Reality* (*local ranking*) by using **Augment** software

Criterion	*Alternetive*	*Ranking*	$\frac{r_p}{\sum_{p=1}^{n} r_p}$	w_{1j}
Aesthetic Impact	*Exposed Concrete*	3	3/(3+1+2+9+8+6)	0.10
	Wood Coated	1	1/(3+1+2+9+8+6)	0.03
	Brick Coated	2	2/(3+1+2+9+8+6)	0.07
	Ceramic Coated	9	9/(3+1+2+9+8+6)	0.31
	Stone Coated	8	8/(3+1+2+9+8+6)	0.28
	Plastered	6	6/(3+1+2+9+8+6)	0.21

Fig. 5.9 Local weights evaluation of the *Aesthetic Impact criterion* by using a spreadsheet

Figure 5.9 shows the calculations carried to achieve local weights $w_{1,j}$ by applying Eq. (5.1) for $i = 1$.

Note that also local weights v_i are computed on the basis of Eq. (3.1) by considering $p = i$.

5.8 AR-AHP Step 4, *Local Consistency Test*

To validate the DMs coherence, a Local Consistency check is performed. In the AR-AHP local consistency test, first two 3D models (*alternatives*) are randomly extracted according to *AR phase* 3 (Sect. 3.5). The extracted *alternatives* are c_4 (*Ceramic Coated*) and c_2(*Wood Coated*). Furthermore, DMs assign a value of $k_{4,2} = 9.5$ to the

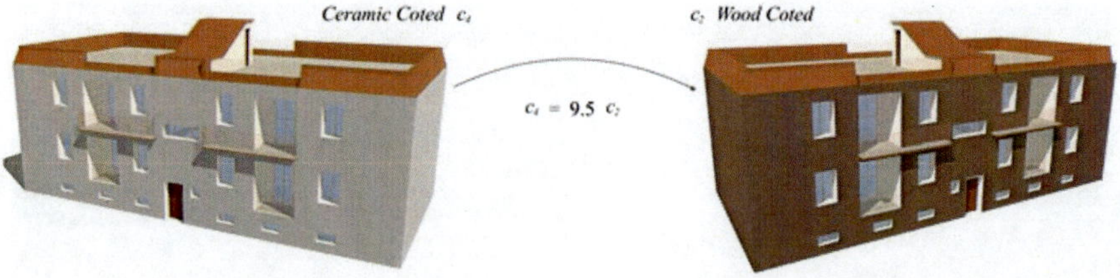

Fig. 5.10 Local Consistency check: best aesthetic impact between c_4 (*Ceramic Coated*) and c_2 (*Wood Coated*)

pair comparison between c_4 and c_2. This value k represents the differences between c_4 and c_2 according to the DMs perception (Fig. 5.10).

In particular, the DMs believe that c_4 has a weight 9.5 times greater than c_2. If in the *AR Phase 2, k* has been consistently applied, k should have a similar value to the ratio between v_4 and v_2. To this aim, by reworking Eq. (3.2), it is possible to verify the Local Consistency:

$$LC_{4,2} = \left| \frac{9.5 - \left(w_{1,4}/w_{1,2}\right)}{w_{1,4}/w_{1,2}} \right|. \tag{5.2}$$

The Local Consistency check is verified since the $LC_{4,2} = 0.056 \leq 0.3$ and this means that the DMs are carrying out the procedure coherently.

Note that AR-AHP Steps 3 and 4 are repeated until all the weights of the decision problem have been obtained. In addition, the AR-AHP possibility allows carrying out the decision making process directly onsite. To provide another example, Fig. 5.11 shows the local ranking of 3D models related to the *Realization Complexity criterion* performed on site.

5.9 AR-AHP Step 5, the Global Weights Evaluation

After obtaining all the local weights of the problem, the tabulated weights related to the *Structure of the Problem* considered are obtained (Fig. 5.12).

The tabulated weights are obtained by organizing in a table all the local weights previously calculated. The tabulated weights show the numerical value associated with the respective *criterion* (weight v_i) or *alternative* (w_{ij}).

By using the tabulated weights, it is possible to calculate the global weights $w_j^{'}$ (AR-AHP step 4) representing the effective preferences of the group of users with regard to the best PCPs selection for the considered case study. In particular,

Fig. 5.11 Ranking of 3D models related to the *Realization Complexity criterion* performed on site

the following equation is used to obtain the global weights in accordance with the classical aggregation of weights by Saaty (2008):

$$w'_j = \sum_{i-1}^{6} v_i \times w_{i,j}, \forall j \in N_2. \tag{5.3}$$

In particular, Eq. (5.3) is applied to i which varies from 1 to 6 to evaluate w'_1, w'_2, w'_3, w'_4, w'_5, w'_6, and reported as follows:

$$w'_1 = \sum_{i=1}^{6} v_i \times w_{i,1} = \left(v_1 \times w_{1,1}\right) + \left(v_2 \times w_{2,1}\right) + \left(v_3 \times w_{3,1}\right) + \left(v_4 \times w_{4,1}\right) + \left(v_5 \times w_{5,1}\right) + \left(v_6 \times w_{6,1}\right);$$

$$w'_2 = \sum_{i=1}^{6} v_i \times w_{i,2} = \left(v_1 \times w_{1,2}\right) + \left(v_2 \times w_{2,2}\right) + \left(v_3 \times w_{3,2}\right) + \left(v_4 \times w_{4,2}\right) + \left(v_5 \times w_{5,2}\right) + \left(v_6 \times w_{6,2}\right);$$

$$w'_3 = \sum_{i=1}^{6} v_i \times w_{i,3} = \left(v_1 \times w_{1,3}\right) + \left(v_2 \times w_{2,3}\right) + \left(v_3 \times w_{3,3}\right) + \left(v_4 \times w_{4,3}\right) + \left(v_5 \times w_{5,3}\right) + \left(v_6 \times w_{6,3}\right);$$

Criterion	v_i	*Alternative*	w_{ij}
Aesthetic Impact	*0.18*	*Exposed Concrete*	0.10
		Wood Coated	0.03
		Brick Coated	0.07
		Ceramic Coated	0.31
		Stone Coated	0.28
		Plastered	0.21
Required Construction Vehicle	*0.03*	*Exposed Concrete*	0.18
		Wood Coated	0.15
		Brick Coated	0.15
		Ceramic Coated	0.18
		Stone Coated	0.14
		Plastered	0.18
Realization Complexity	*0.09*	*Exposed Concrete*	0.30
		Wood Coated	0.13
		Brick Coated	0.03
		Ceramic Coated	0.27
		Stone Coated	0.07
		Plastered	0.20
Thermal Transmittance	*0.24*	*Exposed Concrete*	0.16
		Wood Coated	0.17
		Brick Coated	0.17
		Ceramic Coated	0.17
		Stone Coated	0.17
		Plastered	0.17
Thermal Inertia	*0.26*	*Exposed Concrete*	0.19
		Wood Coated	0.02
		Brick Coated	0.20
		Ceramic Coated	0.19
		Stone Coated	0.20
		Plastered	0.20
Expected Costs	*0.21*	*Exposed Concrete*	0.19
		Wood Coated	0.14
		Brick Coated	0.16
		Ceramic Coated	0.18
		Stone Coated	0.17
		Plastered	0.17

Fig. 5.12 Local weights obtained with the AR-AHP

$$w_4^{'} = \sum_{i=1}^{6} v_i \times w_{i,4} = (v_1 \times w_{1,4}) + (v_2 \times w_{2,4}) + (v_3 \times w_{3,4}) + (v_4 \times w_{4,4}) + (v_5 \times w_{5,4}) + (v_6 \times w_{6,4});$$

$$w_5^{'} = \sum_{i=1}^{6} v_i \times w_{i,5} = (v_1 \times w_{1,5}) + (v_2 \times w_{2,5}) + (v_3 \times w_{3,5}) + (v_4 \times w_{4,5}) + (v_5 \times w_{5,5}) + (v_6 \times w_{6,5});$$

$$w'_6 = \sum_{i=1}^{6} v_i \times w_{i,6} = (v_1 \times w_{1,6}) + (v_2 \times w_{2,6}) + (v_3 \times w_{3,6}) + (v_4 \times w_{4,6}) + (v_5 \times w_{5,6}) + (v_6 \times w_{6,6}).$$

By using the tabulated weights of Fig. 5.12 in the application of Eq. (5.3) for the six i $(i = 1, \ldots, 6)$ the global weights can be calculated and are shown in Fig. 5.13.

In this example, the AR-AHP helped make the following decision: the most effective PCP for the building considered (located in Trani, Italy) is the *Ceramic Coated* by considering *criteria* of aesthetical effect, time and needs of production, installation, technical properties and economic sustainability. Figure 5.14 shows a pie chart where the preferences of the final decision (global weights w'_j) are expressed in percentages. In addition, Fig. 5.15 shows a render of the Building to be restored with the best PCP solution.

Alternative	w'_i	w'_i
Exposed Concrete	w'_1	0.18
Wood Coated	w'_2	0.10
Brick Coated	w'_3	0.15
Ceramic Coated	w'_4	0.21
Stone Coated	w'_5	0.18
Plastered	w'_6	0.19

Fig. 5.13 Global weights obtained with the AR-AHP

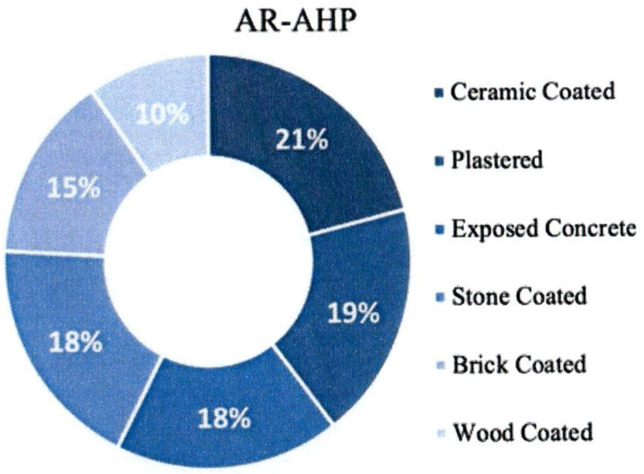

Fig. 5.14 Global weights obtained with the AR-AHP and expressed in percentage

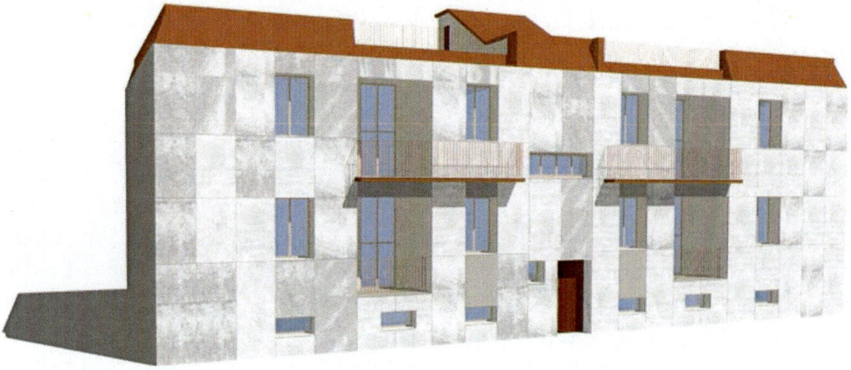

Fig. 5.15 Render of the Building to be restored with the best PCP solution: *Ceramic Coated*

Note that the calibration and validation of an MCA devoted to the identification of the best alternative among a set of possible solutions can be achieved by involving a statistically relevant number of DMs or by performing many applications. In this case, the AR-AHP is performed by all the practitioners (designers) operating in the building renovation of the construction considered. To this aim, the decision can be considered validated for this specific building. In addition, to verify the effectiveness of the proposed AR-AHP, a comparison with the classical AHP is presented in the next section.

5.10 Comparison with the Classical AHP Approach

In this section, we present a comparison with the result of a classical AHP applied to the same PCP selection problem—"to retrofit an economic-popular residential complex located in Trani". In particular, the same decision problem is faced by the same DMs group by using the classical AHP for two purposes. First, it is possible to compare the *results* to verify that the same choice can be obtained by the classical AHP. Second, a useful *procedure* comparison between AR-AHP and AHP can be realized in terms of time, complexity and consistency of the approach.

Note that the results of this comparison are not valid in general, but for specific applications where augmented reality can play a fundamental role (as in the case of the best PCPs selection for building retrofitting). To provide some examples, the augmented reality can be very effective to include in the decision: aesthetics aspects, assembly configurations, specific procedures or maintenance procedures that are complex to be understood. In this context, 3D models displayed in AR can be much more effective than a simple description or images.

The *results* obtained by the two approaches are very similar and only a few differences in the importance of some parameters appear. In particular, the same preferences and final ranking are obtained for the PCP selection as shown in Fig. 5.16. Indeed, the group of practitioners (designers) obtained the following global weights: (i) Ceramic Coated is the most efficient PCP (according to all the methods) thanks to good characteristics in all the criteria and the related global weights expressed in percentage of 21% for both the approaches; (ii) Stone Coated, Exposed concrete and Plastered PCPs have a similar global weight ranging between 17 and 19%; (iii) Brick Coated and Wood Coated PCPs are the worst panels for the considered case study with a weight of 15% and 10/11%, respectively.

The second comparison regards the *procedure* between AR-AHP and AHP.

The main differences have been identified in three fundamental features (Fig. 5.17): (i) the length of the decision making, (ii) the complexity of the approach and (iii) the consistency of evaluation.

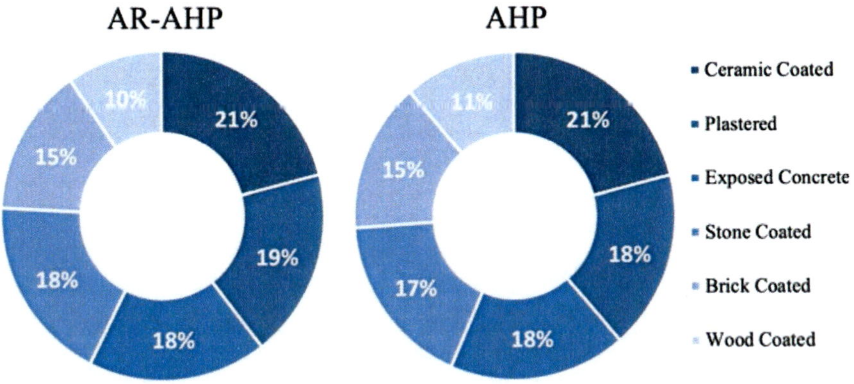

Fig. 5.16 Result comparison between AR-AHP and AHP

Approach	*Average Length (Time)*	*Complexity of the approach*	*Consistency Requirement*
AR-AHP	*15 minutes*	*7 ranking*	*86% at the first attempt*
AHP	*65 minutes*	*Required in total 105 pairs comparison*	*50% at the first attempt*

Fig. 5.17 Procedure comparison between AR-AHP and AHP

(i) The length of the decision making in terms of time is the first important difference. The proposed AR-DM is the faster approach and needs on average 15 min to evaluate seven rankings. On the contrary, the AHP requires a time that can vary from 50 to 80 min to evaluate seven matrices of AHP.

(ii) The complexity of the approach is the second important difference. The AR-AHP is completely based on a visual, fast and intuitive procedure exploiting an augmented reality to support the analysis. The AR environment becomes fundamental to simplify the evaluation process of the DM (particularly for weighting related to Aesthetic Impact or Realization Complexity). Furthermore, the AR-AHP required only 7 rankings to achieve the decision. On the other hand, the classical AHP required in total 105 comparison pairs to get the final result.

(iii) In conclusion, the consistency of the judgments is a typical problem of the AHP methodology (Sangiorgio et al., 2018). Indeed, in AHP applied to the case study, consistency is not immediately reached in over 50% of cases. Consequently, this means that DMs do not have all the necessary information to make a coherent decision (more information needs to be gathered). On the other hand, in the AR-AHP the consistency is almost always immediately reached (at the first attempt for 86% of the evaluation). In particular, in the AR-AHP approach, the AR environment provides all the necessary information to make a coherent decision.

To sum up, the two approaches are effective in terms of achieved results, but the AR-AHP uses a more intuitive AR supported procedure that is able to achieve the same result of the classical AHP in a faster and more consistent way. For the qualitative parameters that require visual information, the DM founds in the AR environment exceptional support.

Note that the consideration of better effectiveness of the AR-AHP in comparison to the AHP is not valid in general but for specific applications where augmented reality can play a fundamental role as in the case of the best PCPs selection for building retrofitting.

References

Boeri, A., & Longo, D. (2011). Energy efficiency in buildings in southern Europe: Challenges and design strategies. *International Journal of Sustainable Development and Planning, 6*(4), 522–536.

Building components and building elements—Thermal resistance and thermal transmittance—Calculation methods, UNI EN ISO 6946:2018 (March 1, 2018).

D.P.R. No. 495 (1992). Regolamento di esecuzione e di attuazione del nuovo codice della strada.

Martiradonna, S., Lombillo, I., & Fatiguso, F. (2019). Performance Monitoring of Refurbished Buildings through Innovative Precast Concrete Modules. *In 2019 ieee international conference on systems, man and cybernetics* (SMC) (pp. 953–957). IEEE.

Martiradonna, S., Fatiguso, F., & Lombillo, I. (2020a) Precast concrete module for structural and energy rehabilitation of reinforced concrete buildings. *Proceedings of REHABEND 2020: Euro-American congress: Construction pathology, rehabilitation technology and heritage management* (pp. 1618–1626).

Martiradonna, S., Fatiguso, F., & Lombillo, I. (2020b) Thermal improvements of existing reinforced concrete buildings by an Innovative Precast Concrete Panel system. *Colloqui.AT.e 2020, New horizons for sustainable architecture*, Italy, 2020. Forthcoming.

Pittau, F., Malighetti, L. E., Iannaccone, G., & Masera, G. (2017). Prefabrication as large-scale efficient strategy for the energy retrofit of the housing stock: An Italian case study. *Procedia Engineering, 180*, 1160–1169.

Saaty, T. L. (2008). Decision making with the analytic hierarchy process. *International Journal of Services Sciences, 1*(1), 83–98.

Sangiorgio, V., Uva, G., & Fatiguso, F. (2018). Optimized AHP to overcome limits in weight calculation: Building performance application. *Journal of Construction Engineering and Management, 144*(2), 04017101.

Sangiorgio, V., Martiradonna, S., Fatiguso, F., & Lombillo, I. (2021). Augmented reality based-decision making (AR-DM) to support multi-criteria analysis in constructions. *Automation in Construction, 124*, 103567

Chapter 6
User Reporting and AHP to Investigate the Perception and Social Acceptance of Wind Energy

Abstract The investigation of the stakeholder's *perception* on a large scale, it is a complex problem characterized by multiple actors with often conflicting values and views. The modern technologies combined with multi-criteria decision analysis (MCDA) can offer fundamental support to engage a large number of stakeholders. On the other hand, a structured *User Reporting* procedure is required to limit and overcome the typical drawbacks of data acquired by users. This chapter shows how the application of a *User Reporting* supported MCDA can investigate the perception and the social acceptance of wind energy in cities of southern Italy by overcoming the classical drawbacks of data acquired by users. **Firstly**, the *stakeholders* are identified and classified in clusters, in order to have homogeneous samples of *Users* and limits conflicting views. **Secondly**, the *technological tool* (survey administration software) is selected to reach a large number of *stakeholders* and squire a statistically significant number of data. **Thirdly**, the data are analyzed by using the *Analytic Hierarchy Process* to obtain a statistical graph of the user perception. Finally, the results are used in order to draw suitable guidelines useful to the *Wind Energy Companies* aimed at minimizing the environmental and social impact of wind power plants.

Keywords Stakeholder's perception · Social acceptance · Case study · Wind energy · User Reporting · Stakeholders analysis · Questionnaires · Flowchart · Data analysis · Architectural engineering · Building production

This Chapter proposes a *User Reporting* supported MCA to reach and interview a large number of stakeholders and investigate the perception and the social acceptance of wind energy in cities in southern Italy, close to wind farms.

The goal of the MCA regards the evaluation of the stakeholder's *perception* (*MCA Goal C* defined in Sect. 1.4). The purpose of *User Reporting* is to support the decision making involving a large number of DMs (**first case "*User Reporting supported decision making*"**, theory explained in Chap. 4). In addition, the MCA used is the AHP (explained from a theoretical point of view in Chap. 2).

First, we give a brief description of the problem in Sect. 6.1.

Second, we use the AHP to investigate the stakeholder's perception and discuss in detail all the involved parameters involved in Sect. 6.2.

V. Sangiorgio et al., *New Approaches for Multi-Criteria Analysis in Building Constructions*, https://doi.org/10.1007/978-3-030-83875-1_6

Third, we apply the eight steps of *User Reporting* explained in Chap. 4 to this case (Sects. 6.3–6.11).

6.1 The Social Perception of Wind Energy

Wind energy power has seen rapid development worldwide in recent decades. There is evidence of largely acquired benefits such as competitiveness, sustainability, lower energy costs and energy independence for this type of energy. On the other hand, several controversies have arisen in recent years in terms of social impact and acceptance. This is observed in a particular way in Europe (Jobert et al., 2007) with some peculiar differences in Italy where the existence of a well-developed national market of renewable energy and several potential perspectives for future developments still leaves unsolved questions mainly related to the social sphere (Caporale et al., 2020). Indeed, although wind turbines installation represents a sustainable alternative to fossil fuels, this technology can make a considerable cumulative impact on the landscape, especially in terms of visual impact, noise and climatic impacts on wildlife. As a consequence of such controversies, the importance of the social perception of wind energy in the correct management and operation of wind farms is fundamental as emphasized in the study of Caporale and De Lucia (2015). In that study, the authors show the high impact on the local community of the costs incurred to install and manage this type of renewable energy.

In this context, *User Reporting* supported decision making can be very effective to investigate the parameters involved in the social acceptance of wind energy in southern Italy. Many parameters can be considered in order to understand which are the critical aspects that make wind energy one of the most discussed renewable technology in the social sphere (Thayer & Freeman, 1987). In particular, the purpose of *User Reporting* is to involve a large set of user/DMs to quantify the importance of the parameters involved in the social acceptance of wind energy such as aesthetic impact, environmental sustainability, economic sustainability, etc. In addition, the MCA applied in conjunction with *User Reporting* is the AHP. The area investigated includes the cities close to wind farms located in Apulia (Italy) in the neighbourhood of Bari and Foggia.

Note that in this application the *calibration* is achieved by involving a statistically significant number of DMs.

6.2 The AHP Integrated in *User Reporting*

In this application of *User Reporting* (**first case**) the preliminary step is *Structuring of the problem* being investigated according to the AHP. A literature review is conducted in order to identify and corroborate all the defined *criteria* and *sub-criteria* involved in the social perception of wind energy. An exhaustive discussion of such parameters is proposed in Caporale et al. (2020). The resulting main criteria regarding the perception of wind farms may be summarized as follows: *aesthetic impact*, *environmental sustainability*, *economic sustainability*, *functional efficiency*, *noise* and *inadequacy of institutions*. In turn, each criterion can be decomposed into additional sub-criteria in a hierarchical scheme. In total, in this decision problem, we considered 6 criteria and 25 sub-criteria (divided into two levels of sub-criteria).

(i) The *Aesthetic impact* is a fundamental parameter in perception, impact and desirability of wind installations. Landscape and visual disturbance of facilities is an important disadvantage of the wind farms, in particular for residents living in neighbouring cities. Indeed, the wind turbines could be seen from as far as 30 km during the day under clear sky conditions (Bishop, 2002). People living close-by to a wind farm and who are familiar with the original landscape are more affected by its impact. Generally, this impact is influenced by turbine colours, size and density per unit of land area. Consequently, this criterion can be decomposed into five sub-criteria such as *number of turbines*, *turbines distance*, *turbines dimension*, *turbine colour* and *location* (that can be *sea*, *plain* or *hill*).

(ii) The perception of *environmental sustainability* is a complex issue that goes beyond the widespread concept of renewable energy generation reducing the dependence on fossil fuels of countries. Indeed, during the construction operation and dismantling of wind farms, various alterations and risks can arise and affect the local bio-system. In particular, the soil surface alteration and soil erosion are causes of the local eco-balance modification and agricultural production alteration. This issue could arise during the foundation excavation and road construction of a wind farm. Consequently, these operations can jeopardize wildlife safety. In the review of Dai et al. (2015) it is demonstrated that the mortality and disturbance risks that wind turbines induce on birds are severe. However, different land contexts and predator removal rates make difficult the exact estimation of a bird fatality rate. In addition, during the dismantling, other environmental problems could arise if all the connected operations are not properly designed (Arnett et al., 2011; Foote, 2010; Liechti et al., 2013; Marsh, 2007). To investigate the social perception of the *environmental sustainability* it is possible to take into account five sub-criteria, including *management impact*, *implementation impact*, *dismantling impact*, *fauna alteration* and *agricultural production alteration*.

(iii) The perception of *economic sustainability* of a wind investment project generally plays a great role in terms of market and management structure. Indeed, economic benefits can be perceived by nearby communities such as sharing of project profits, new employment opportunities, tourism and other socio-economic effects to boost the local economy (Burguillo & Rio, 2008). From the expert point of view, the Italian National Association of Wind Energy considers that the most significant benefit of wind energy production is the reduction of national unity price (named PUN) (Caporale et al. 2020). Consequently, to the PUN reduction, a saving of about 6.6 billion Euros is forecasted during the period 2018–2030. On the other hand, high installation costs appear to be a relevant obstacle for wind investments. The International Renewable Energy Agency estimates that after the year 2010, installation costs of a new onshore wind farm is between 1850 and 2100 USD/kW in Europe (Caporale et al. 2020). Wind turbines account for 64 to 84% of total installed costs, with grid connection costs, construction costs and other costs 'rounding' the balance. Beyond the installation costs, also maintenance and dismantling are important expenses to be considered. Consequently, the economic sustainability criterion can be decomposed into four sub-criteria *profit, maintenance costs, implementation costs, dismantling costs.*

(iv) The *functional efficiency* is another parameter that must be considered in the desirability of this technology (Herbert et al., 2010). The reliability of a wind farm is high during the first years of operation, while it drastically decreases over time beyond economic viability. The amount of energy production and average daily operation depends on the limits of the technology and the intensity and frequency of the winds, respectively. On the other hand, wind farm lifetime and efficiency are also related to human actions such as the proper maintenance: periodical cleaning of blades, use of proper heat treatment of the blade rope for strengthening, employ of adequate lubricants to avoid high temperature in the gearbox, maintaining of adequate stock of high-quality gear oil for periodical replacement. In total, the *functional efficiency* in the social perception of wind energy can be related to three sub-criteria comprising *useful life, amount of energy production and average daily operation.*

(v) The *noisiness* of the turbines can be of three different typologies. The noise related to the non-aerodynamic instabilities and unstable airflows, the noise generated by the interaction of wind turbine blades and atmospheric turbulence and the mechanical noise that comes from the turbine's internal gears, the generator and other auxiliary parts (Pantazopoulou, 2010; Pedersen, 2011). Governments and medical institutions usually recommend a minimum distance between the location of wind farms and the surrounding buildings because high frequency noise can induce sleep disturbance and hearing loss, headaches, irritability, fatigue, arteries constriction, immune system weakening, annoyance and dissatisfaction (Dai et al., 2015). Starting from this bibliographic investigation, it is possible to define three sub-criteria involved in the perception of the *noisiness* of wind farms: *turbine distance, turbine dimension, number of turbines.*

(vi) The last criterion regards the *Inadequacy of the institutions*. In some regions of southern Italy, the mistrust of local institutions for a correct management of public land, incentives and great works can affect the positive perception of a wind farm project. Some authors demonstrated that the adequacy of institutions is the key point in the building of trust across stakeholders and in the social acceptance by local communities (Stigka et al., 2014).

The initial support of local and national governments is an important precondition for the revision of the land use plan. Usually, inconsistency at provincial and municipal level planning turned out in objections for several administrative stakeholders at different levels of authority.

Moreover, when the residents are not involved in the project, they can produce a negative opinion by perceiving the wind energy as an external private interest passing over the interests of the local community. The shared economic interest in wind power exploitation would result as the main driver for public collaboration and acceptance.

On the other hand, it is not clear in the related literature how much important the citizens perceive the *Inadequacy of the institutions* in comparison with the other defined criteria. In this application, in order to clarify the significance of this criterion, two sub-criteria are considered: *misinformation* and *no transparency of public procurement*.

A complete overview of the involved criteria and sub-criteria is shown in Fig. 6.1.

Note that the other steps of the AHP (*Local weights evaluation*, *Local consistency test* and *Global weights evaluation*) are integrated with the user reporting and discussed in the following subsections.

6.3 Identification of Stakeholders

The first step is to set the *User Reporting* can start with a brainstorming during which the decision managers (in this case the researchers who intend to investigate the users' perception) develop a list of potential stakeholders:

- *Citizens* of the cities near wind farms;
- *Local Public Authorities,* who may be interested in the outcome of the project but without being active participants;
- *Researchers*, which in this case are the decision managers directly involved in the analysis and validation of results;
- *Wind Energy Companies* who are interested in the results of the analysis to limit the social impact of wind farms with adequate planning.

At the end of the brainstorming, the stakeholders are confirmed and classified according to the *Power/interest grid* given in Fig. 6.2:

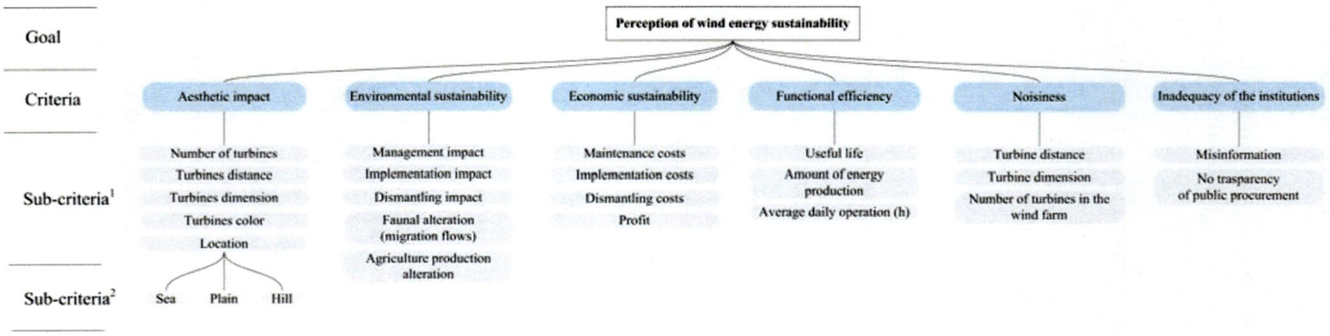

Fig. 6.1 The *structuring of the problem in a hierarchy*: criteria and sub-criteria to assess the social perception of wind farms

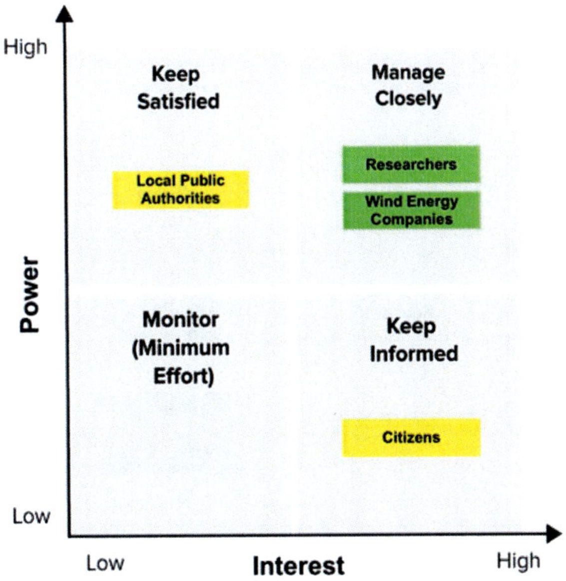

Fig. 6.2 *Power/interest grid* to support the stakeholder's classification in the case of the social perception of wind energy

High power, highly interested people are Researchers and Wind Energy Companies;
High power, less interested people are individuated in Local Public Authorities;
Low power, highly interested people are the Citizens of the cities near wind farms;
Low power, less interested people are not present in this case study.

6.4 Definition of Users

Once the stakeholders are identified, they can be classified according to the *Numerosity/Expertise grid* in order to understand the quality and quantity of data that can be acquired from every group. In particular, the classification developed is explained as follows:

High numerosity,high expertise stakeholders are identified in the *Citizens* since in this application, the objective is the evaluation of the inhabitants' perception;
High numerosity,low expertise are not present in this application;
Low numerosity,high expertise stakeholders are identified in two potential classes: *Researchers* and *Wind Energy Companies*. Such classes are not the users of the reporting since their perception can be affected by economic or scientific interest in the wind energy production;

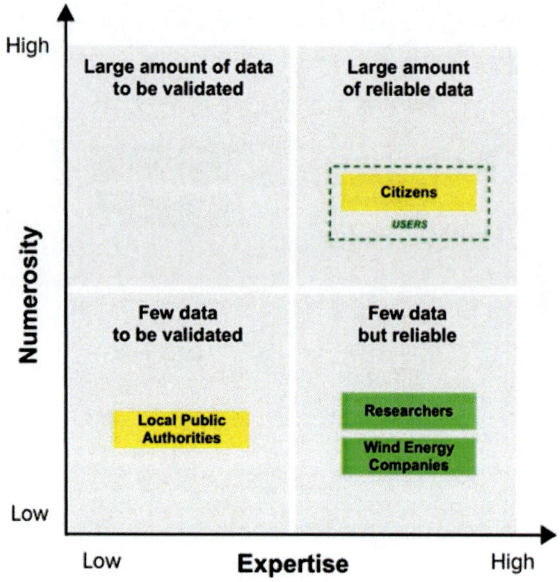

Fig. 6.3 *Numerosity/expertise grid* to support the identification of users

Low numerosity,low expertise can be represented by *Local Public Authorities* that cannot provide useful information because also in this case they could have a direct interest in the development of wind farms. Consequently, their perception can be distorted by professional interests.

From this classification and after the decision managers discussion, among the stakeholders, the potential users (of the *User Reporting*) are identified and indicated in the *Numerosity/Expertise grid* showed in Fig. 6.3. In particular, the *users* are identified in the *Citizens* because they can provide their own perception of wind energy without having a direct interest in the realization of wind farms. The *Researchers* and *Wind Energy Companies* can validate the acquired data and exploit the result of the *User Reporting*, respectively. The *Local Public Authorities* need only to be informed about the outcome of the investigation.

6.5 Selection of Technological Tools to Support the Data Acquisition

In the investigation of wind energy perception, the technological tool can be an effective instrument to reach a large number of users (DMs). Indeed, it is necessary to acquire a large set of data to evaluate the perception of the citizens defined in the *Numerosity/Expertise grid*.

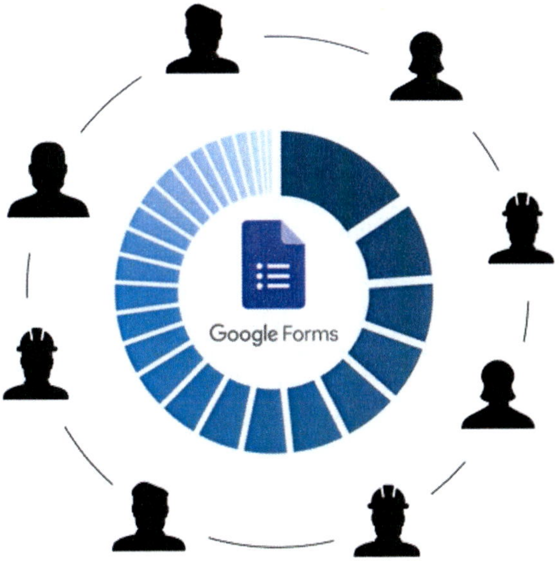

Fig. 6.4 Web based Docs Editors suite of Google is chosen as a technology tool to address support the *User Reporting* (*Google Forms*, web based Docs Editors suite offered by Google)

Among the available technological tools, survey administration software can be used in order to submit a questionnaire linked to a multi-criteria approach. In this case, the *Google Forms* software (Fig. 6.4) is used (web based Docs Editors suite offered by Google). The application allows users to create and edit surveys online while collaborating with other users in real-time. The collected information can be automatically entered into a spreadsheet.

6.6 Creation of Questionnaires

The creation of the questionnaires is related to the information required to perform the AHP and evaluate the social perception of wind farms. In particular, for every investigated user, eight *Judgment Matrices* need to be filled with a total of 51 questions. In addition, the matrices and the related questions required for the decision problem (structured in Fig. 6.1) are listed as follows.

In the level of "Sub-Criteria^2" only one matrix and relative questions are required:

(i) A matrix 3×3 and 3 related questions to investigate the sub-criteria^2 related to the parameter *location*.

In the level of "Sub-Criteria[1]" six matrices with the related questions are necessary:

(ii) A matrix 5 × 5 and 10 related questions to investigate the sub-criteria related to the criterion *aesthetic impact*;
(iii) A matrix 5 × 5 and 10 related questions to investigate the sub-criteria related to the criterion *environmental sustainability*;
(iv) A matrix 4 × 4 and 6 related questions to investigate the sub-criteria related to the criterion *economic sustainability*;
(v) A matrix 3 × 3 and 3 related questions to investigate the sub-criteria related to the criterion *functional efficiency*;
(vi) A matrix 3 × 3 and 3 related questions to investigate the sub-criteria related to the criterion *noisiness*;
(vii) A matrix 2 × 2 and 1 related question to investigate the sub-criteria related to the criterion *inadequacy of institutions*.

In the level of "Criteria[1]" one matrix is necessary:

(viii) A matrix 6 × 6 and 15 related questions to investigate the criteria related to the goal *perception of wind energy sustainability*.

Note that the number of necessary questions to obtain the judgments required for each *Judgment Matrix* can be calculated according to Eq. (2.1) and specified by the number of investigated parameters in the matrix (see Chap. 2).

To provide an example of a question, let us consider the criterion of the *Aesthetic Impact*, in order to identify the pair comparison between the two sub-criteria *Number of Turbines* and *Dimension of Turbines*. The question and some of the possible answers obtained with *Google Forms* are shown in Fig. 6.5.

Note that in this example, the answers are not directly related to the numerical judgment, but the values of the relative *Judgment Matrix* need to be identified by using the verbal scales and the associated numerical values of Saaty' scale (Table 2.1) (**Equal** importance 1, **Moderate more** importance of one over another 3, **Strong more** importance 5, **Very strong more** importance 7, **Extreme** importance 9. And reciprocal values including **Moderate less** importance of one over another 1/3, **Strong less** importance 1/5, **Very strong less** importance 1/7, **Extreme less** importance 1/9 and reciprocal values).

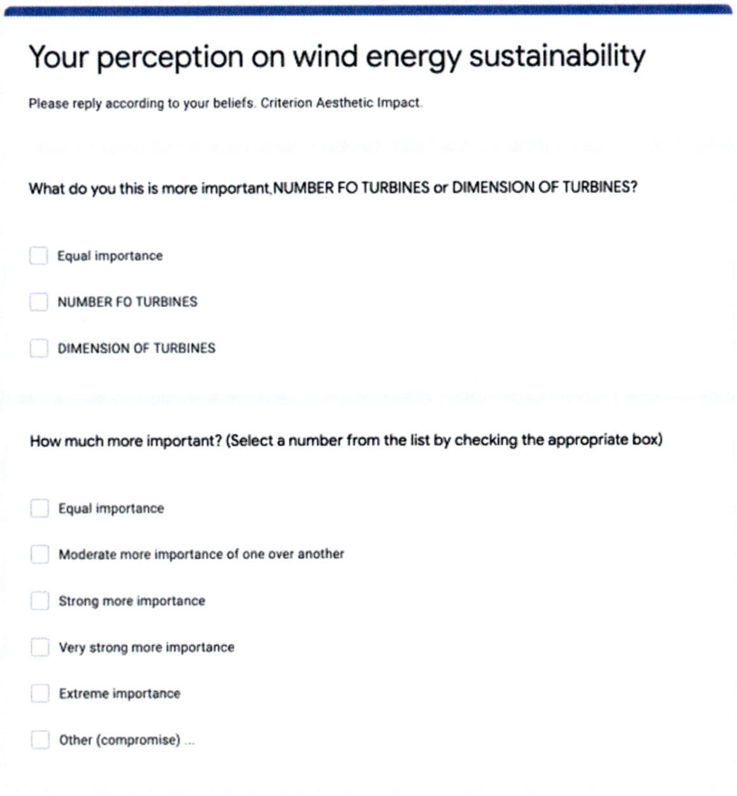

Your perception on wind energy sustainability

Please reply according to your beliefs. Criterion Aesthetic Impact.

What do you this is more important,NUMBER FO TURBINES or DIMENSION OF TURBINES?

- [ ] Equal importance
- [ ] NUMBER FO TURBINES
- [ ] DIMENSION OF TURBINES

How much more important? (Select a number from the list by checking the appropriate box)

- [ ] Equal importance
- [ ] Moderate more importance of one over another
- [ ] Strong more importance
- [ ] Very strong more importance
- [ ] Extreme importance
- [ ] Other (compromise) ...

Fig. 6.5 Example of the questionnaire: question regarding the importance of sub-criteria *Number of Turbines* and *Dimension of Turbines* with respect to the *Aesthetic Impact* criterion

6.7 Definition of the Flowchart Describing the *User Reporting*

In this subsection, the flowchart of the procedure is defined for the case of the social perception of wind energy (Fig. 6.6). The UML activity diagram shows in the columns the clusters of stakeholders and technologies used to perform the actions of the *User Reporting*. In particular, the flow starts from the *Researchers* that create the questionnaire. Subsequently, the Google form is used to reach a large number of citizens.

Once the information is acquired by the Google form, the data obtained are processed and analyzed by the *Researchers*. After *Researchers* verify the consistency of the AHP matrices, the weights are stored and employed to obtain a statistical graph of the user perception to investigate the distribution of the global rankings obtained for each user and display the median of the sample. Finally, the User Reporting

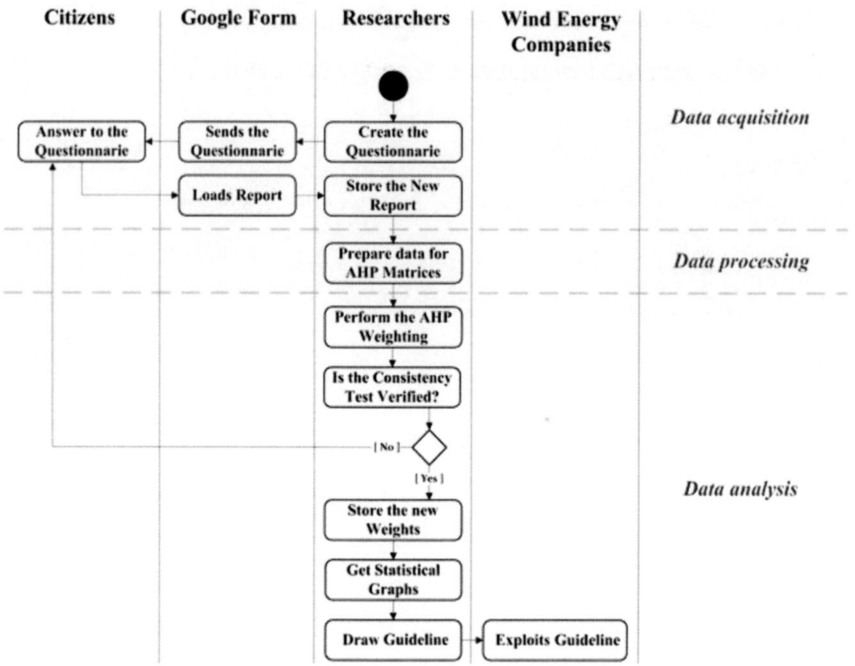

Fig. 6.6 Flowchart describing the *User Reporting* (**First case**) for the wind energy perception described by using UML framework

results allow drawing suitable guidelines useful to the *Wind Energy Companies* with the aim of minimizing the environmental and social impact of wind power plants.

6.8 Effective Data Acquisition of the Citizen Perception

In this step, the operative data acquisition is performed through the *User Reporting*. The *Researchers* distribute the questionnaire through social networks or by direct communication with the inhabitants of cities near the wind farms.

To provide an example, the set of information regarding the perception of the *Aesthetic Impact* is reported by user #31 (where #31 is the id code assigned to the considered user) citizen of Foggia (Apulia, Italy) by answering 10 out of 51 questions. In particular, the criterion *Aesthetic Impact* includes the following sub-criteria: the number of turbines, the distance between turbines, the dimension, colour and location of turbines. Consequently, the 10 questions that allow to obtain the verbal judgments of user #31 regarding the perception of the *Aesthetic Impact* are listed as follows:

Answer 1: *Number of turbines* is **strongly less** important than *Turbines distance*;
Answer 2: *Number of turbines* is **moderately more** important than *Turbines dimension*;
Answer 3: *Number of turbines* is **very strongly more** important than *Turbines colour*;
Answer 4: *Number of turbines* is **very strongly more** important than *Turbines location*;
Answer 5: *Turbines distance* is **moderately more** important than *Turbines dimension*;
Answer 6: *Turbines distance* is **extremely more** important than *Turbines colour*;
Answer 7: *Turbines distance* is **extremely more** important than *Turbines location*;
Answer 8: *Turbines dimension* is **strongly more** important than *Turbines colour*;
Answer 9: *Turbines dimension* is **strongly more** important than *Turbines location*;
Answer 10: *Turbines colour* is **equally importance** with *Turbines location.*

6.9 Processing of the Citizen Perception Data

Once the questionnaire is completed by the user, the data are prepared to fill the *Judgment Matrices*. Since in this application the questionnaire is qualitative, the acquired answers are converted in numerical values according to the verbal scales of Saaty.

To provide an example the set of information regarding the perception of the *Aesthetic Impact* is reported by user #31 citizen of Foggia are converted in numerical values as follows:

Answer 1: **strongly less** = 1/5;
Answer 2: **moderately more** = 3;
Answer 3: **very strongly more** = 7;
Answer 4: **very strongly more** = 7;
Answer 5: **moderately more** = 3;
Answer 6: **extremely more** = 9;
Answer 7: **extremely more** = 9;
Answer 8: **strongly more** = 5;
Answer 9: **strongly more** = 5;
Answer 10: **equally importance** = 1.

Note that the same questionnaire could be provided directly by asking the numerical value. In this case, the questionnaire and the results would appear as showed in Fig. 6.7.

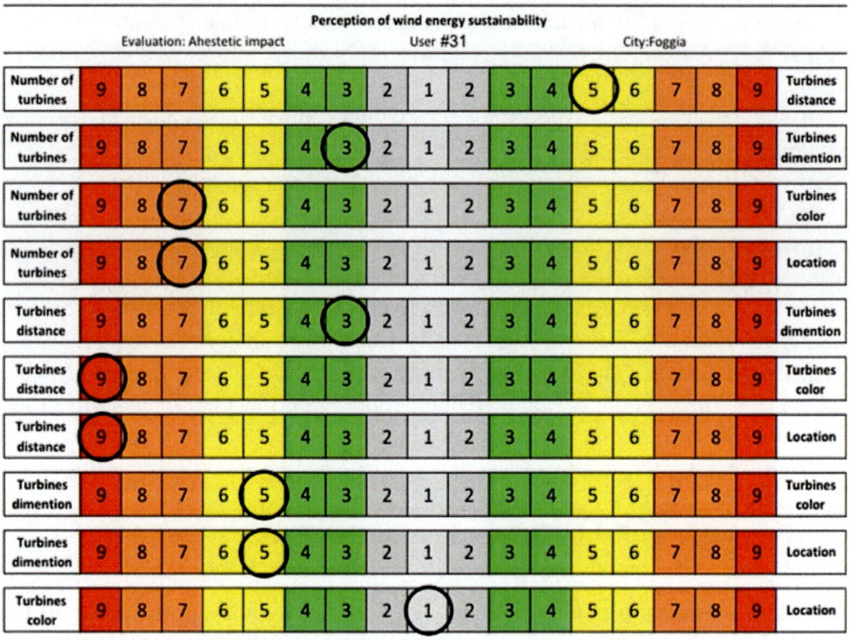

Fig. 6.7 Example of the questionnaire asking directly the numerical values, questions regarding the importance of sub-criteria with respect to the *Aesthetic Impact* criterion

Fig. 6.8 *Judgment Matrix* to evaluate the importance of sub-criteria with respect to the *Aesthetic Impact* criterion according to the perception of user #31 (DM)

Aesthetic Impact User #31		*Nt*	*Td*	*Tm*	*Tc*	*Tl*
Number of turbines	*(Nt)*	1.0	0.3	3.0	7.0	7.0
Turbines distance	*(Td)*	3.0	1.0	3.0	9.0	9.0
Turbines dimension	*(Tm)*	0.3	0.3	1.0	5.0	5.0
Turbines color	*(Tc)*	0.1	0.1	0.2	1.0	1.0
Turbines location	*(Tl)*	0.1	0.1	0.2	1.0	1.0

When the numerical values related to the answer to the 10 questions are obtained, these can be used to fill the relative *Judgment Matrix* of the *Aesthetic Impact* for the corresponding user #31 (Fig. 6.8).

6.10 Data Analysis and Guideline for the Wind Energy

In the last step of the *User Reporting* procedure the local weights are evaluated, the consistency test is verified and the global weights are evaluated according to the classical AHP procedure.

Note that the weight evaluation is carried out for each criterion and for each involved user.

In the following, the example of the weight evaluation for the *Aesthetic Impact* according to the perception of user #31 is reported.

By solving the eigenvector problem according to Eq. (2.2), or alternatively by using the simplified procedure of Eqs. (2.3 and 2.4), the local weights are calculated (and expressed in percentages %). In addition, the values of CR are calculated according to Eqs. (2.5 and 2.6) and possibly by the simplified procedure of Eqs. (2.7 and 2.8) (the value of RI is $n = 5$, see Table 2.2). In this case, the consistency limits are satisfied since $\mathrm{CR} = 0.05 < 0.1$ and consequently, the weights can be considered reliable. If the weights were not consistent, the user would be asked to answer this part of the questionnaire again (the 10 questions regarding the *Aesthetic Impact*). Figure 6.9 shows the *Judgment matrix* of the *Aesthetic Impact* (user #31) including the values of CR and the local weights expressed in percentages (%).

When all the involved users answered all questions consistently, the global weights and the final ranking of the parameters can be obtained for every user. The global weight of every sub-criterion can be computed by multiplying the local weight of the considered sub-criterion with the local weight of the relative criterion according to the classical aggregation principle of Saaty.

To provide an example, if the local weight of the Criterion *Aesthetic Impact* is 0.2 (20%) and the local weight of the sub-criterion *number of turbines* is 0.281 (28.1%), the global weight of the sub-criterion *number of turbines* is equal to $0.2 \times 0.281 = 0.056(5.6\%)$. By comparing all the global weights of the considered parameters, it is possible to obtain the global ranking. The global ranking represents the classification from the most important parameter to the less important one according to the user perception.

When a statistically significant number of users' perceptions is acquired and the global weights are evaluated, a statistical boxplot of the citizens' perceptions can be identified. Figure 6.10 shows the boxplot of the citizen perception where ranking = 1 denotes the most important parameter in the common citizen's perception influencing the wind energy desirability. In particular, the boxes represent the distribution of the rankings and the black line inside the boxes denotes the median of the sample.

Aesthetic Impact User #31		*Nt*	*Td*	*Tm*	*Tc*	*Tl*	*CR*	Local weights (%)
Number of turbines	*(Nt)*	1.0	0.3	3.0	7.0	7.0		28.1
Turbines distance	*(Td)*	3.0	1.0	3.0	9.0	9.0		47.3
Turbines dimension	*(Tm)*	0.3	0.3	1.0	5.0	5.0	0.05	16.4
Turbines color	*(Tc)*	0.1	0.1	0.2	1.0	1.0		4.1
Turbines location	*(Tl)*	0.1	0.1	0.2	1.0	1.0		4.1

Fig. 6.9 *Judgment Matrix*, CR and local weights of the *Aesthetic Impact* according to the perception of user #31

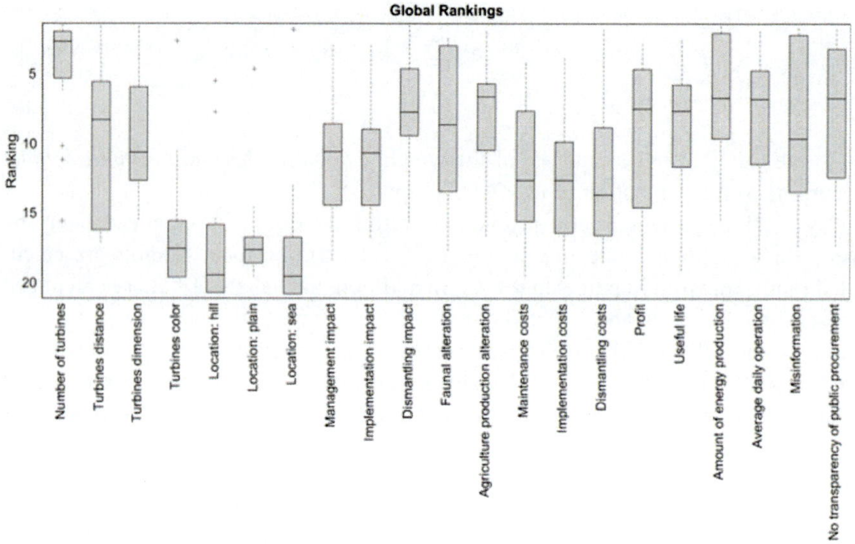

Fig. 6.10 Boxplot of the citizen perception, global ranking of investigated parameters

In addition, the box contains all the sub-criteria within the 25th and 75th percentile of the population. This is a measure of the dispersion of the perceptions. The dashed line encompasses all the sub-criteria considered as outliers. According to the median of the sample shown in the Boxplot of Fig. 6.10 (representing the global ranking of investigated parameters), the most important parameter affecting the desirability of a wind farm (for the involved citizens) is the *number of turbines.* Other fundamental parameters are the *agriculture production alteration*, the low *amount of energy production*, the *limited average daily operability* and the *no transparency of public procurement.*

Such results can be very effective for local administrations and wind energy companies in order to reduce the social impact in the design of new wind farms. In particular, some **guidelines** can be drawn up from the results of this study. Some suggestions regard the design of the turbines that can be improved by considering that the dimension does not significantly influence the aesthetic impact. Other recommendations concern the development of information campaigns in order to reduce the mistaken beliefs of citizens. For instance, it is possible to enlighten the impact of wind energy on the agriculture industry in comparison with other typologies of energy production.

6.11 Conclusion of the Social Perception of Wind Energy Investigation

The *User Reporting* results emphasize that the criteria environmental sustainability (i.e. faunal, agricultural alteration and dismantling impact), the functional efficiency (i.e. the amount of produced energy) and the inadequacy of institutions (i.e. misinformation and distrust) exhibit the highest value of importance in terms of the perception of wind energy desirability. Furthermore, the AHP global ranking results obtained also help to understand social acceptance of wind energy in terms of willingness to accept social costs. In particular, the costs attributed to maintenance, implementation and dismantling are significantly less important compared to the profits. As a consequence, the users are willing to accept such costs if they could rely on an adequate economic incentive.

Note that, the output of such an application is a ranking of the parameters representing the perception of the users. To this aim, a calibration/validation has been performed by involving a *statistically significant number of DMs*. Moreover, it is worth noting that results were validated for the case of Apulia region, more precisely for the city neighbourhood of Bari and Foggia.

References

Arnett, E. B., Huso, M. M., Schirmacher, M. R., & Hayes, J. P. (2011). Altering turbine speed reduces bat mortality at wind-energy facilities. *Frontiers in Ecology and the Environment, 9*(4), 209–214.

Bishop, I. D. (2002). Determination of thresholds of visual impact: The case of wind turbines. *Environment and Planning B: Planning and Design, 29*, 707–718. https://doi.org/10.1068/b12854

Burguillo, M., & del Rio, P. (2008). Assessing the impact of renewable energy deployment on local sustainability: Towards a theoretical framework. *Renewable and Sustainable Energy Reviews, 12*, 1325–1344. https://doi.org/10.1016/j.rser.2007.03.004

Caporale, D., & De Lucia, C. (2015). Social acceptance of on-shore wind energy in Apulia Region (Southern Italy). *Renewable and Sustainable Energy Reviews, 52*, 1378–1390. https://doi.org/10.1016/j.rser.2015.07.183

Caporale, D., Sangiorgio, V., Amodio, A., & De Lucia, C. (2020). Multi-criteria and focus group analysis for social acceptance of wind energy. *Energy Policy, 140*, 111387.

Dai, K., Bergot, A., Liang, C., Xiang, W., & Huang, Z. (2015). Environmental issues associated with wind energy—A review. *Renewable Energy, 75*, 911–921. https://doi.org/10.1016/j.renene.2014.10.074

Foote, R. (2010). The wind is blowing the right way for birds. *Renewable Energy Focus, 11*(2), 40–42.

Herbert, G. M. J., Iniyan, S., & Goic, R. (2010). Performance, reliability and failure analysis of wind farm in a developing country. *Renewable Energy, 35*, 2739–2751. https://doi.org/10.1016/

Chapter 7
User Reporting and Condition Ratings to Support Building Maintenance and Diagnostics

Abstract The use of *Key Performance Indicators* (KPIs), specific *Condition Ratings* (*Cr*) or risk indexes obtained with multi-criteria analysis (MCA) is widespread in the construction sector and can be very effective for the **building maintenance** and **diagnostics**. In addition, the synergy of MCA with the *User Reporting* ensures a large-scale application of the indexes, exploiting data acquired by the users, supported by modern technological tools. This chapter shows a practical application of the *User Reporting* applied to the identification of maintenance and diagnostics intervention priority in coastal cities of the Valencia region. Preliminary, an *Analytic Hierarchy Process* has been developed to define suitable *Condition Ratings* for the building condition assessment. Afterwards, *Condition Ratings* are applied and the *User Reporting* allows to reach a large quantity of reliable data. Technological tools such as an application software (APP) for smart devices and a suitable *Quality Detection Platform* (QDP) support the large data acquisition. The APP also allow the users to send the report to the QDP that automatically store the information. The QDP is a useful tool for the *Building technicians* that can display the reports and the calculated *Condition Ratings*. The procedure allows to identify the priority of intervention in the Valencian Coastline.

Keywords Building maintenance · Building diagnostics · Architectural engineering · Building production · Condition ratings · User Reporting · Case study · Technological tools · Stakeholders analysis · Questionnaires · Flowchart · Data analysis

This chapter shows a practical application of *User Reporting* applied to building maintenance and diagnostics. In this case, the goal of *User Reporting* is to perform a large-scale analysis exploiting the results of a preliminarily developed MCA model (**Second Case—"*User Reporting supported large-scale analysis*"**, theory explained in Chap. 4). In particular, the AHP is used to define suitable *condition ratings* that have been already validated for the building condition assessment (*MCA Goal A* defined in Sect. 1.4). More precisely, the problem regards the building degradation analysis and the identification of maintenance priorities in coastal cities of the region of Valencia (Spain).

V. Sangiorgio et al., *New Approaches for Multi-Criteria Analysis in Building Constructions*, https://doi.org/10.1007/978-3-030-83875-1_7

Note that in this chapter, the objective is not focused on the use of the MCA but is devoted to show the potential of *User Reporting* to perform a large-scale building degradation analysis by exploiting the resulting concise indexes (named *condition ratings*, Cr) of a preliminary MCA model.

The necessary information to be acquired by the *User Reporting* regards all the parameters to calculate the *condition ratings* of the buildings investigated.

To this aim, first, we present a concise overview of the preliminary MCA model (carried out by AHP) (Sects. 7.1, 7.2 and 7.3) and second, the *User Reporting* process is explained in detail (Sects. 7.4, 7.5, 7.6, 7.7, 7.8, 7.9, 7.10, 7.11 and 7.12).

7.1 The Case of Valencia

The Coast of Valencia is 440 km long. It is bathed by the Mediterranean Sea and contains 65 townships. In the last half century, the coastline has become a popular area for vacation homes and as a result, it has undergone a profound metamorphosis (Menero & Garrido, 2011). The most widespread constructions are the multi-story holiday-apartment buildings, which are more numerous than those used by permanent residents. The building typology used in these areas is the apartment block, although there are also many terraced houses. The structure of the buildings is usually made of precast joist slabs and Reinforced Concrete (RC) frames. The RC structure is often visible thanks to the presence of open-plan ground floors. In addition, the thickness of the concrete cover is very thin and most of the time even below 2 cm (Moreno et al., 2015). These buildings are therefore highly exposed to the aggressive marine environment (Adam et al., 2016; Moradian et al., 2015) . Consequently, the corrosion of the reinforcement by the sea atmosphere is one of the main causes of the degradation and criticalities of these RC buildings (Angst et al., 2009; Jin et al., 2016; Pradhan, 2014). Figure 7.1 shows the typical constructions of the Valencian

Fig. 7.1 Typical construction and damages related to the aggressive marine atmosphere

coast and recurrent damages related to the aggressive marine atmosphere. To this aim, these buildings need constant monitoring and evaluation of existing damage conditions.

7.2 The Preliminary AHP to Evaluate the Building Degradation

This section briefly presents the preliminary AHP methodology consisting of five steps applied to the building degradation analysis: (i) *structuring of the problem* in a hierarchy, (ii) *local weights evaluation*, (iii) *local consistency test*, (iv) *global weights evaluation* and, (v) *sensitivity analysis*. The goal is to achieve a weighting of all the involved *criteria* useful for a fast-visual inspection of the damages and pathologies of RC buildings subject to the aggressive marine environment.

The **first step** of AHP consists in *structuring of the problem* of the structural degradation assessment of RC buildings. Starting from the goal (large-scale RC*building degradation* analysis), the problem is structured in the set C of criteria i with $Cardinality\ (C) = m = 4$:

1 = *Damaged component*, i.e. the importance of the damaged component;
2 = *Damage extension*, i.e. the (percent) damage extension in relation to the component dimension;
3 = *Damage severity*, i.e. the gravity of the damage;
4 – *Component Position*, i.e. the position of the component described by the floor in which it is located.

> **Note** that in this case, the AHP is developed to define suitable *condition ratings* and consequently, a rigorous mathematical formalization is required.

In addition, for each *criterion* $i \in C$, a set B_i of *alternatives* $b_{ij} \in B_i$ with $Cardinality\ (B_i) = \alpha_i$ is defined. Figure 7.2 shows the hierarchical scheme of the problem structure. Once the problem has been defined, a set of experts in the field of RC building construction degradation has been involved to develop the other steps of the AHP.

In particular, the AHP helps to obtain the weights that are defined as follows:

v_i is the weight associated with each *criterion* $i \in C$;
w_{ij} is the weight associated with each *alternative* $b_{ij} \in B_i$ with respect to a *criterion* $i \in C$.

Figure 7.2 shows the association of weight v_i to the ith *criterion* for $i = 1,\ldots,m$ and the weights w_{ij} to the jth *alternative* for $j = 1,\ldots,\alpha_i$. In this figure, $m = 4$, $\alpha_1 = 22$, $\alpha_2 = 10$, $\alpha_3 = 25$ and $\alpha_4 = 3$.

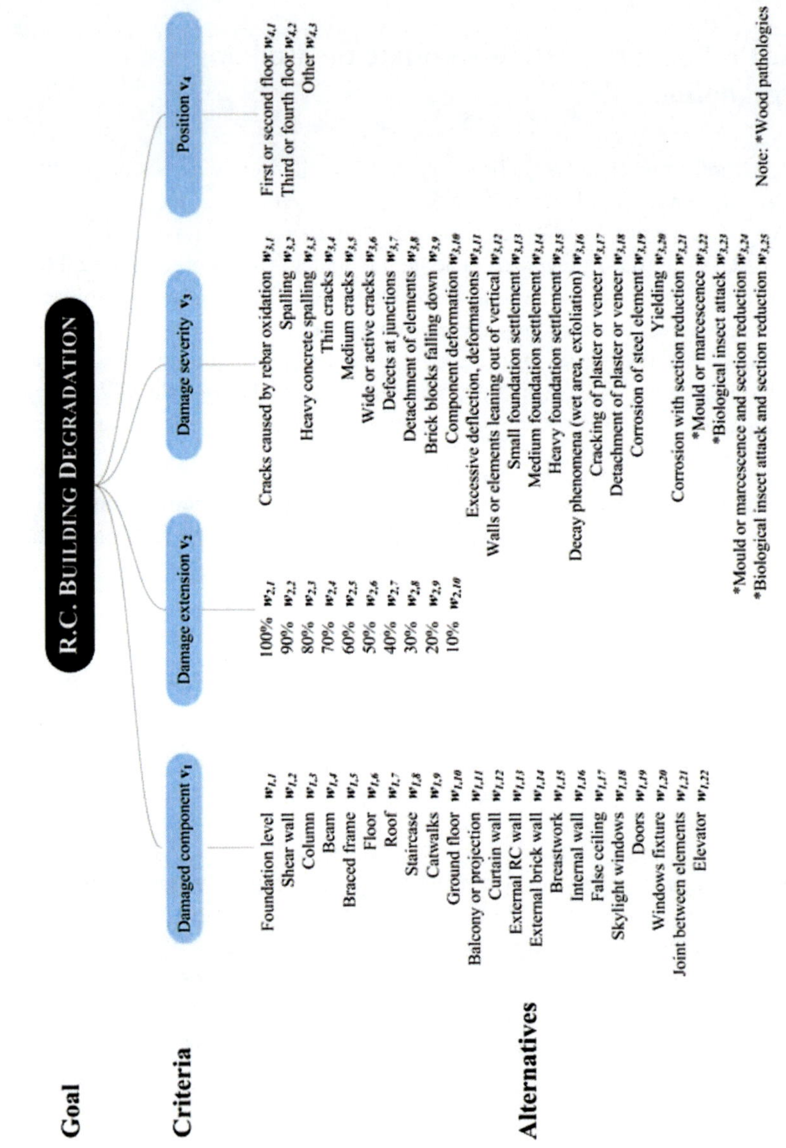

Fig. 7.2 Hierarchical structure of the problem for the RC building degradation analysis

Note that in the *damaged severity criterion*, for each type of damage, 3 increasing severity levels are defined. For example: thin cracks ($w_{3,4}$), medium cracks ($w_{3,5}$) and wide or active cracks ($w_{3,6}$) are the levels corresponding to the Cracks damage.

In the **second and third steps,** the AHP is performed by obtaining five *Judgment Matrices*: four matrices to identify the weights w_{ij} of the *alternatives* and one matrix to identify the weight v_i of each *criterion.* The resulting local weights are shown in Table 7.1.

Note that after the five steps of the AHP, the weights were also *calibrated* according to the Tuutti model (1982) to consider the damage evolution of reinforcements' corrosion. In addition, such results are *validated* by comparison with a similar approach of the related literature widely used in South America.

The next sections show how in the fourth step of the AHP the o local weights obtained are used to develop *condition ratings* that can be employed along with thc *User Reporting* approach in order to assess a large number of constructions.

Table 7.1 Tabulated weights: weights of criteria and alternatives obtained with the AHP for the RC building degradation

Weights of *criteria* and *alternatives* obtained with the AHP													
v_1	$w_{1,1}$	$w_{1,2}$	$w_{1,3}$	$w_{1,4}$	$w_{1,5}$	$w_{1,6}$	$w_{1,7}$	$w_{1,8}$	$w_{1,9}$	$w_{1,10}$	$w_{1,11}$	$w_{1,12}$	$w_{1,13}$
0.1	10	9.1	7.9	6.1	6.1	6.1	6.1	6.1	6.1	6.1	5.5	2.4	2.2
v_2	$w_{2,1}$	$w_{2,2}$	$w_{2,3}$	$w_{2,4}$	$w_{2,5}$	$w_{2,6}$	$w_{2,7}$	$w_{2,8}$	$w_{2,9}$	$w_{2,10}$			
0.43	10.0	8.3	6.2	5.5	4.5	3.6	2.8	2.1	1.4	0.7			
v_3	$w_{3,1}$	$w_{3,2}$	$w_{3,3}$	$w_{3,4}$	$w_{3,5}$	$w_{3,6}$	$w_{3,7}$	$w_{3,8}$	$w_{3,9}$	$w_{3,30}$	$w_{3,11}$	$w_{3,12}$	$w_{3,13}$
0.44	2.1	3.9	8.8	2.1	4.4	9.1	2.1	3.9	3.9	2.4	8.3	9.1	5.1
v_4	$w_{4,1}$	$w_{4,2}$	$w_{4,3}$	$w_{4,4}$	$w_{4,5}$	$w_{4,6}$	$w_{4,7}$	$w_{4,8}$	$w_{4,9}$	$w_{4,10}$			
0.03	10	6.2	1.1										
Weights of *criteria* and *alternatives* obtained with the AHP													
v_1	$w_{1,14}$	$w_{1,15}$	$w_{1,16}$	$w_{1,17}$	$w_{1,18}$	$w_{1,19}$	$w_{1,20}$	$w_{1,21}$	$w_{1,22}$				
0.1	2.2	1.7	1.7	1.7	1	1	1	1	1				
v_2													
0.43													
v_3	$w_{3,14}$	$w_{3,15}$	$w_{3,16}$	$w_{3,17}$	$w_{3,18}$	$w_{3,19}$	$w_{3,20}$	$w_{3,21}$	$w_{3,22}$	$w_{1,23}$	$w_{3,24}$	$w_{3,25}$	
0.44	5.1	10.0	2.0	2.1	3.9	0.7	3.5	3.5	0.7	0.7	3.5	3.5	
v_4													
0.03													

7.3 Definition of Concise Indices (Condition Ratings)

Once the results of the MCA model are obtained, it is possible to define three *condition ratings* and set a *User Reporting* procedure for the large-scale application **exploiting the tabulated weights obtained**. Such operation coincides with the **fourth step** of the AHP (*global weights evaluation*).

In particular, the indexes help to quantify the damage of a single component, of a class of components or of the whole building, by acquiring the following information through a visual inspection report: *Damaged component, Damage extension, Damage severity* and *Component Position* according to the structure of the AHP problem.

Now, we present the mathematical formalization used to obtain *condition ratings* for the building degradation analysis.

First of all, the building is described by a set of classes of components $F = \{f = 1,\ldots, N_F\}$ (shear walls, columns, beams, etc.) and a set of components $E = \{e = 1,\ldots, N_E\}$.

Let us assume that H reports are received for a building with the associated information regarding the surveyed damages. For each visual inspection h (with $h = 1, \ldots, H$), it is possible to assign a set of *alternatives* to each *criterion*. Hence, it is possible to define the sets S^h (with $h = 1, \ldots, H$) that collect the *alternatives*, denoted as b_{ij}^h, that are associated with each *criterion* i:

$$S^h = \left\{b_{ij}^h | b_{ij}^h \in B_i, i \in C\right\} \quad \text{with} \quad h = 1, \ldots, H$$

The single damage index D_h with $h = 1, \ldots, H$ is given by the following formula (Sangiorgio et al., 2018a, 2018b):

$$D_h = \sum_{i=1}^{m} v_i \times w_{ij}^h \quad \text{with} \quad j \in \{1, \ldots, \alpha_i\} \tag{7.1}$$

where each i is associated with only one j and where v_i is the weight of the i-th *criterion* and w_{ij}^h is the weight assigned to $b_{ij}^h \in S^h$.

Let us consider a generic component e and the subset of Δ reports related to e. Once the single damage D_h has been obtained, the first *Criticality Condition Rating* (Cr_e), devoted to quantifying the degradation of component e, is defined as a function of the values of D_h as follows:

$$\mathrm{Cr}_e = D_{\max}\left(1 + \frac{\sum_{h=1}^{\Delta} D_h - D_{\max}}{2 * \sum_{h=1}^{\Delta} D_h}\right) \tag{7.2}$$

where D_{max} is the maximum value obtained for D_h with $h = 1, \ldots, \Delta$.

Equation (7.2) is inspired by the heuristic index of Castro (1994) and is improved and validated to be used with the tabulated weights of the AHP in Sangiorgio et al. (2018c).

In order to clarify how to employ such indexes, the following example simplifies the comprehension of the procedure. Under the assumption that two damages are surveyed in a specific component (*Balcony or projection*) of a building and consequently reported by the *User Reporting*, the following indexes can be evaluated:

Damage 1 characteristics.

Damaged component:	*Balcony or projection*	$w_{1,11} = 5.5$
Damage extension:	*80%*	$w_{2,3} = 6.2$
Damage severity:	*Heavy concrete spalling*	$w_{3,3} = 8.8$
Component Position:	*First floor*	$w_{4,1} = 10$

By using Eq. (7.1) the single damage 1 (*Heavy concrete spalling*) of the considered *Balcony or projection* is evaluated as $D_1 = 7.4$ (in a range from 0 to 10).

Damage 2 characteristics.

Damaged component:	*Balcony or projection*	$w_{1,11} = 5.5$
Damage extension:	*40%*	$w_{2,7} = 2.8$
Damage severity:	*Brick blocks falling down*	$w_{3,9} = 3.9$
Component Position:	*First floor*	$w_{4,1} = 10$

By Eq. (7.1), the single damage 2 (*Brick blocks falling down*) of the considered *Balcony or projection* is evaluated as $D_2 = 3.8$ (in a range from 0 to 10).

Consequently, the combination of the damages D_1 and D_2 yields the *component condition rating* of the considered *Balcony or projection* according to Eq. (7.2). Such rating is equal to Cr = 8.9 (normalized to 10), where 10 corresponds to critical damage.

Figure 7.3 provides an example of the visual inspections and related damages D_1, D_2 and Cr_e.

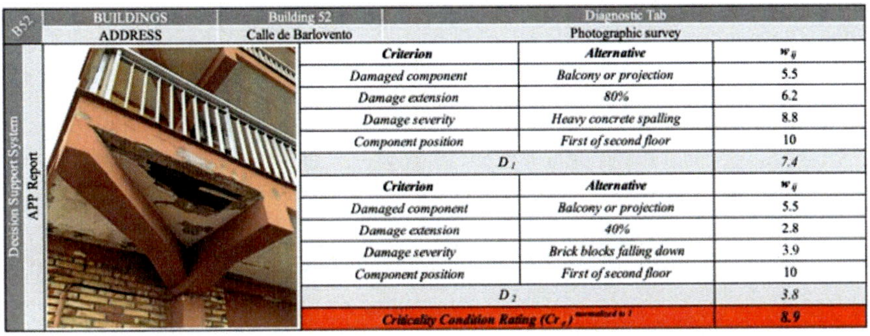

Fig. 7.3 Typical construction and damages related to the aggressive marine atmosphere

Moreover, by acquiring a large set of data about the building, it is possible to evaluate the conservation state of specific classes of components and of the whole construction.

These characteristics can be evaluated by aggregating the values of single damages in **two additional indexes**:

(1) The *Component Condition Rating* (CCr) to evaluate the state of conservation of the building class of component f;
(2) The *Building Condition Rating* (BCr) to evaluate the state of conservation of the whole building considering all its components e.

More precisely, these indices CCr and BCr are computed as follows:

(1) The *component condition rating* of class f (CCr_f) is calculated by aggregating the values of Cr_e representing the *criticality condition rating* of component e belonging to the same f-th class (obtained by Eq. 7.2):

$$\mathrm{CCr}_f = \frac{\sum_e Cr_e}{\gamma_k * S_{\mathrm{tot}}}, \tag{7.3}$$

where S_{tot} is the building total area and γ_k is the approximated evaluation of the number of components considered per square meter defined in Sangiorgio et al. (2019a);

(2) The *Building Condition Rating* (BCr) is calculated by aggregating the values of Cr_e associated with all the elements belonging to the same building using the following expression:

$$\mathrm{BCr} = \frac{\sum_{e=1}^{N_E} \mathrm{Cr}_e}{\gamma_b * S_{\mathrm{tot}}} \tag{7.4}$$

where S_{tot} is the building total area and γ_b is the approximated evaluation of the number of components per square meter, evaluated for the specific RC building typology (Sangiorgio et al., 2019a).

Note that the proposed indexes (Cr, CCr and BCr) are validated by comparison with other consolidated approaches. Moreover, a suitable sensitivity analysis has been performed to verify the robustness of the results. Such validation and calibration are described in detail in Sangiorgio et al. (2019a, 2019b).

7.4 The User Reporting to Set a Large-Scale Maintenance and Diagnostics

At this point, a large set of data acquisitions is required to acquire any existing damage inspections for every considered building. In this application, the goal of *User Reporting* is to provide a visual inspection by including the following information in each report: *Damaged component, Damage extension, Damage severity* and *Component Position.*

The following eight steps of *User Reporting* are applied to this case study. These steps were already described in Chap. 4. They are as follows:

(1) *Identification of the stakeholders*;
(2) *Definition of the Users*;
(3) *Selection of technological tools* to support the acquisition (e.g. Smart Devices);
(4) *Creation of questionnaires and user guideline* (or any choice experiments);
(5) *Definition of the flowchart* (describing all the User Reporting process);
(6) *Effective Data acquisition*;
(7) *Data processing and validation*;
(8) *Data analysis.*

In the next subsections, the eight steps are applied to the case of the degradation analysis in cities located on the coast of Valencia. We describe each step in a section.

7.5 Identification of the Stakeholders in the Coast of Valencia

The first step can start by brainstorming during which the following list of potential stakeholders is gathered:

- *Not expert Users*, stated as inhabitants of the building investigated *without competence* in the construction field;
- *Expert Users,* defined as inhabitants of the building investigated *with competence* in the construction field;
- *Building Technicians*, identified as the people responsible for monitoring, management and safety of these buildings;
- *Local Public Authorities,* who may be interested in the outcome of the project but without being active participants;
- *Other Local Residents* who are neighbours, i.e. inhabitants of the coastal cities of Valencia, but not users of the buildings investigated.

At the end of the brainstorming session, the stakeholders are confirmed and classified according to the *Power/interest grid* (Fig. 7.4) as follows.

High power, highly interested people are identified in *Building Technicians* and *Expert Users* of the Buildings.

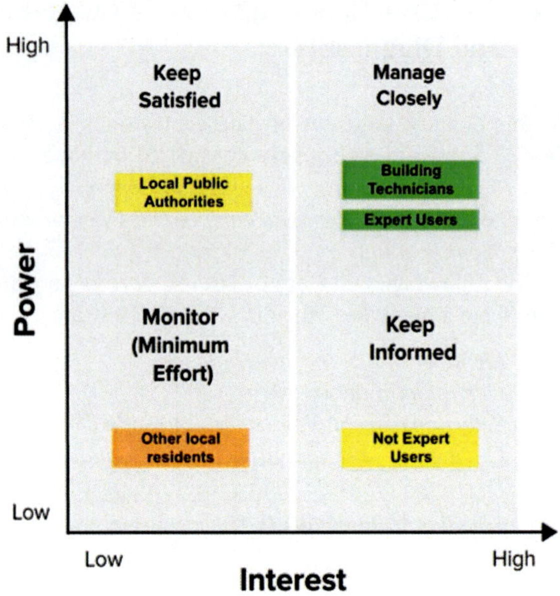

Fig. 7.4 *Power/interest grid* to support the stakeholders' classification in the case of building degradation analysis on the coast of Valencia

High power, less interested people are recognised in *Local Public Authorities.*
Low power, highly interested people are identified in *Not expert Users* of the buildings.
Low power, less interested people are represented by *Other Local Residents.*

7.6 Definition of Building Users in the Coast of Valencia

Once the stakeholders are classified, among these ones the potential Users (of the *User Reporting*) can be identified according to the *Numerosity/Expertise grid.*

Note that, this was already discussed in Chap. 4. This step is fundamental to identify the typology and quality of information the Users can provide during the reporting.

The potential Users are defined and classified according to the *Numerosity/Expertise grid* (Fig. 7.5) as follows.

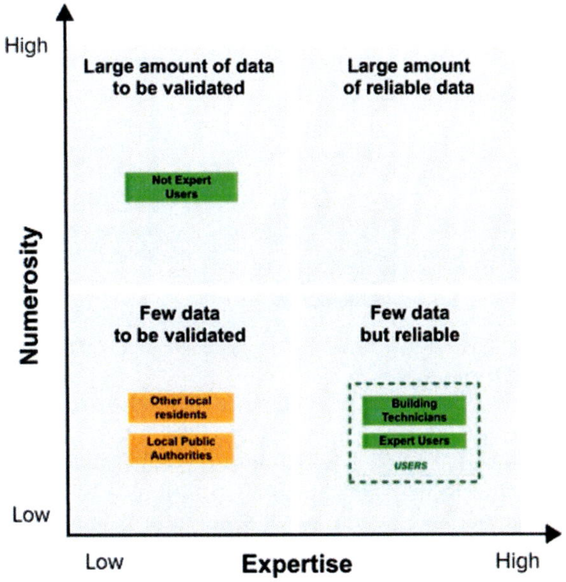

Fig. 7.5 *Numerosity/expertise grid* to support the stakeholders' classification in the case of building degradation analysis on the coast of Valencia

High numerosity, high expertise users are not present in this case study.
High numerosity, low expertise users can be represented by the *Not Expert Users* of the building who are able to provide a large set of data to be validated.
Low numerosity, high expertise users are identified in two potential classes: the *Building technicians* and *Expert Users*. The number of these users is not high, but they can provide reliable information.
Low numerosity, low expertise users can be represented by *Other Local Residents* that may not be able to provide useful information because they do not live or work in the building considered.

7.7 Selection of Technological Tools to Support the Case of the Valencian Building Degradation Analysis

The technological tools are selected starting from the analysis of the *Numerosity/Expertise grid*. Indeed, from the analysis and classification of stakeholders it is apparent that the technological tools can provide the following advantages:

(i) Improving the *numerosity* of reported data by involving a large number of *Expert* and *Not Expert Users*,

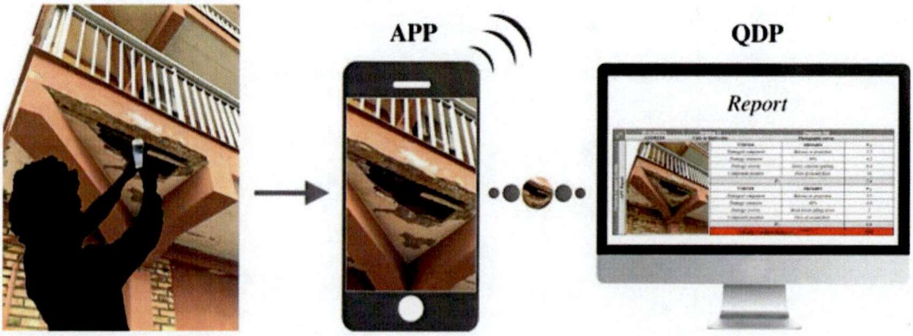

Fig. 7.6 The scheme of the connection between the technological tools of the User Reporting (APP for smart devices and QDP)

(ii) Ensuring a good quality of acquired information to improve the user *expertise.* To this aim, technological support is provided for guiding the user in the acquisition of photographic surveys and information regarding *Damaged component, Damage extension, Damage severity* and *Component Position.*

(iii) Allowing the *Building technicians* to display the reported data to validate or revise the information.

In order to provide the listed advantages, among the available technological tools (discussed in Sect. 4.3) an APP for mobile devices and a suitable Web Based Platform named *Quality Detection Platform* (QDP) are chosen to support the User Reporting in Valencia, (Fig. 7.6).

The operators of the APP are identified among the *Expert Users* and *Not Expert Users* of the building. On the other hand, the operators of the QDP are the *Building technicians.*

The APP allows the *Expert Users* and *Not Expert Users* the acquisition of photographs for inspection and provide additional information related to the critical physical conditions by using a guided questionnaire. The APP also allows sending the report to the QDP that automatically stores the information. The QDP is a useful tool for the *Building technicians* that can display the reports and validate or revise related information. Once validated, data can be prepared for the building degradation analysis by using the *Condition Ratings.*

Figure 7.7 shows how the selected technological tools are used in the last three (operational) steps of the *User Reporting*: *Data Acquisition, Data* Processing and *Data Analysis.*

Note that this procedure helps to overcome the typical problems of retrieving information and systematic visual surveys and monitoring of a large number of buildings.

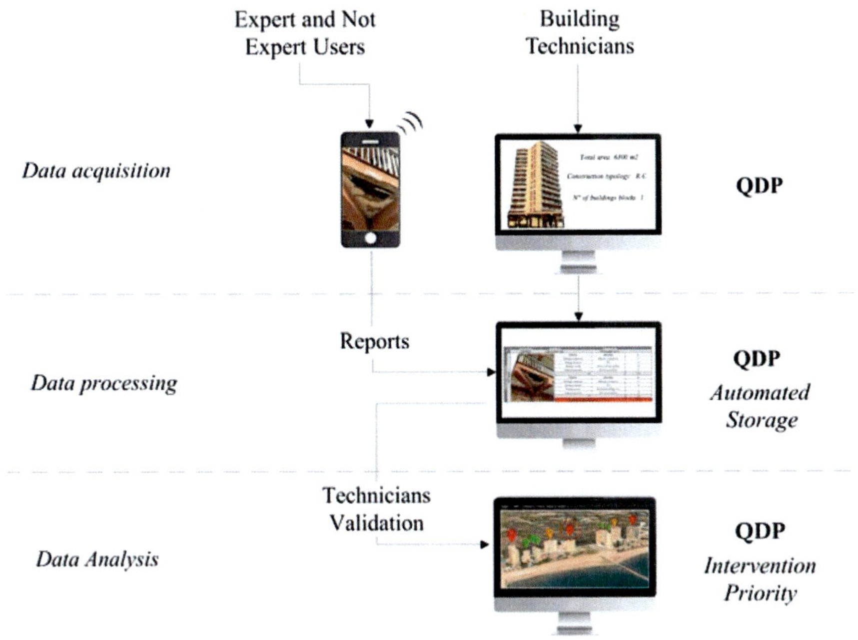

Fig. 7.7 The technological tools in the three phases of *User Reporting*

7.8 Creation of Questionnaires and Guidelines

The APP can be employed by the users of the building to report their visual inspection, i.e. photos and other useful information about the building's damages. It is based on two main operations: photographs and compilation of a guided questionnaire. The photographs enable the users to send images about the criticalities equipped with labels indicating the localization and the date of the picture. Moreover, to complete the description, two structured alternative questionnaires are proposed to guide the user and correctly identify additional data involved in the reported damage (Fig. 7.8). A questionnaire is designed for non-professional people (*Not Expert Users*) and another questionnaire is designed for people with specific competencies in the building field (*Expert Users*).

Whoever the user is, the questionnaire is designed to acquire the following information: (i) *Component Position* identifying the criticality location (floor, room, surrounding elements, building component and height above walking surface); (ii) the *Damage extension* expressed in percentage; (iii) *Damaged component* (such as beam, column, floor, etc.) and the *Damage severity* indicating the type of damage. It is worth noting that each answer of the user opens a subsequent set of possible answers for the next step.

On the basis of the user answers, the APP suggests how to proceed with the photographic inspection or perform other reports in a guided procedure. To provide

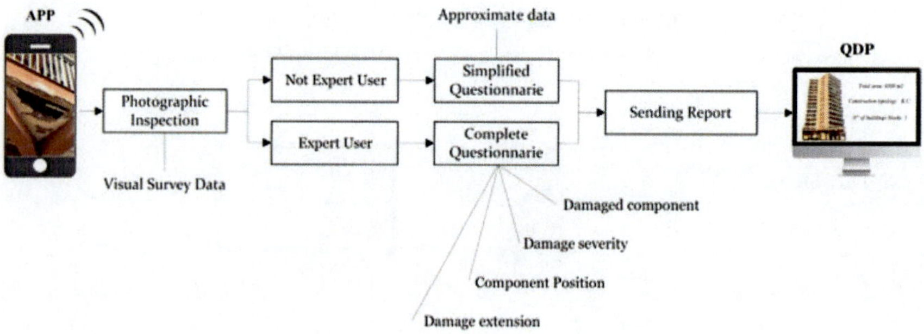

Fig. 7.8 The technological tools and the two structured questionnaires for *Not Expert Users* and *Expert Users*

an example, if the user detects a rebar concrete oxidation in an indoor environment, the APP suggests investigating the possible presence of wet areas. In this way, the user is guided in performing an effective photographic inspection. Figure 7.9 shows an extract of the two questionnaires designed for *Not Expert Users* and *Expert Users.*

7.9 Definition of the Flowchart Describing the *User Reporting*

At this point, stakeholders, users, technologies and questionnaires are defined, consequently, a useful flowchart of the procedure can be sketched to clarify the whole process of the *User Reporting* for the proposed example (Fig. 7.10). The UML activity diagram reports in the columns the users, technologies and other stakeholders that perform the actions listed in the corresponding columns. In the row, the diagram indicated the three operational steps of *User Reporting*: *Data Acquisition, Data Processing* and *Data Analysis* (steps 6, 7, 8 respectively). Note that in the *Data Acquisition* step, the APP selects the expert or non-expert questionnaires on the basis of the typology of users (expert or non-expert). In addition, in the *Data Processing* step the building technicians check all the user reports (both from expert and non-expert) in order to validate or correct the answers of the users. Finally, in the *Data Analysis* step, the condition ratings are evaluated and the intervention priority is defined. In addition, if some KPIs overcome specific threshold values (described in detail 7.11 in the next subsection), the building technicians are asked to perform a prompt on-site survey.

Note that the operational steps 6, 7, 8 are discussed in detail in the following sections.

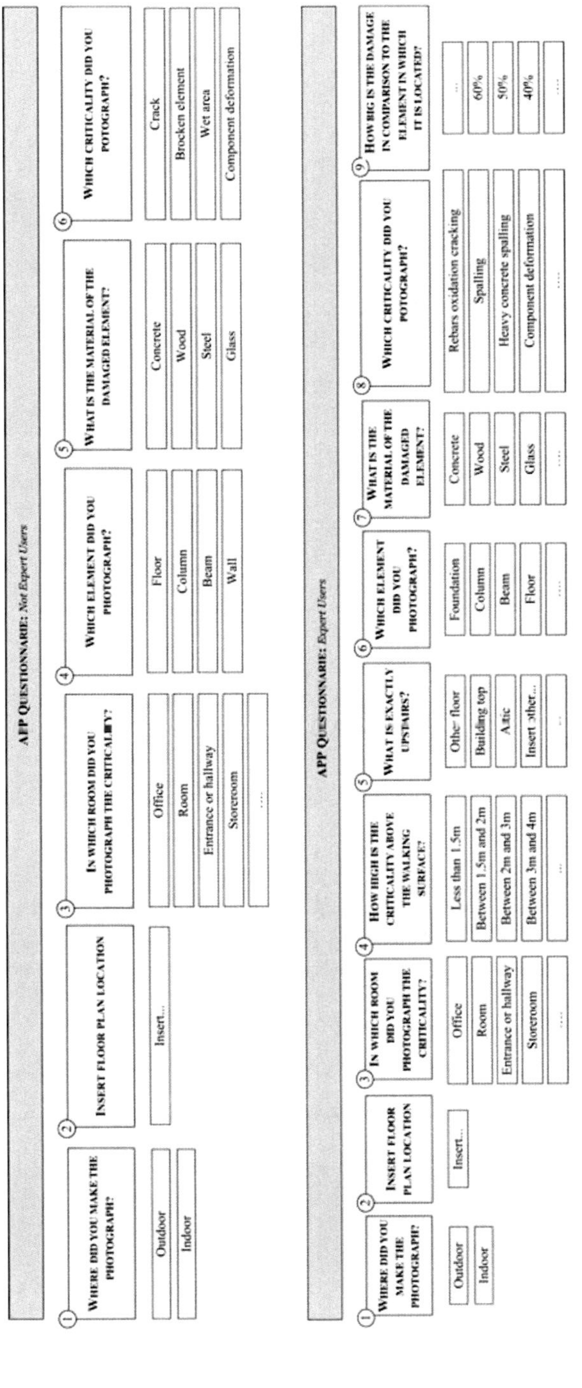

Fig. 7.9 Extract of the two questionnaires: *Not Expert Users* and *Expert Users*

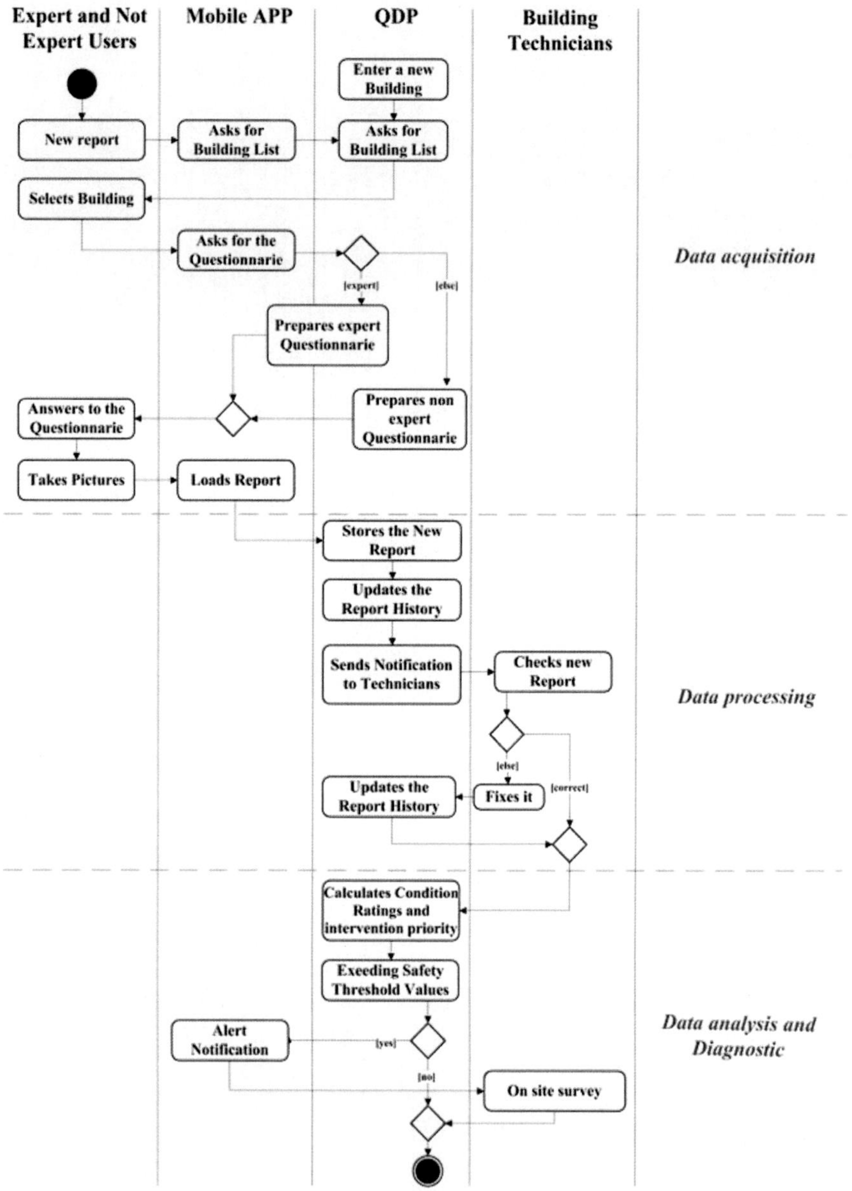

Fig. 7.10 Flowchart procedure described by using UML framework of the *User Reporting* (**second case**)

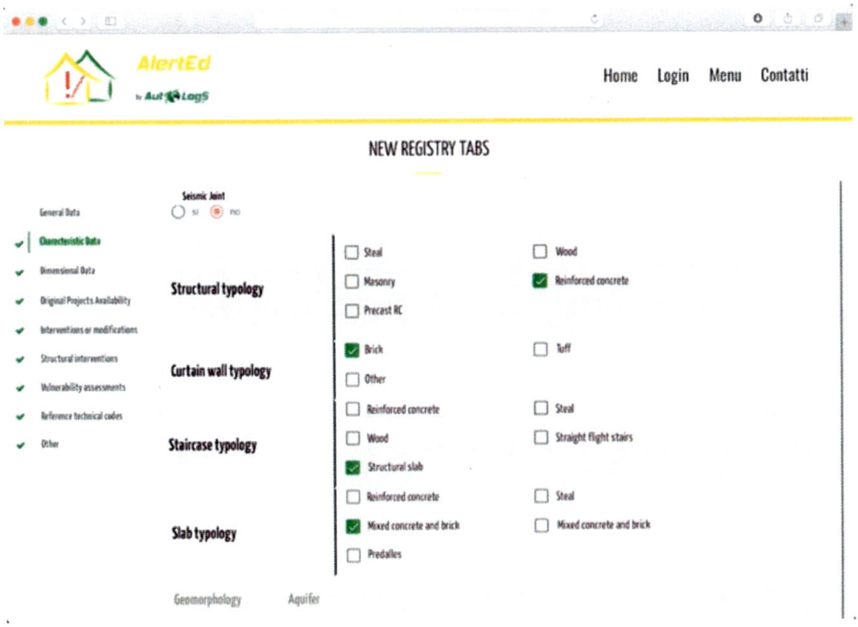

Fig. 7.11 Suitable registry tabs implemented in the QDP to include registry data of every building

7.10 Effective Data Acquisition of the Valencian Building

In this step, the operative data acquisition through the *User Reporting* is performed. First of all, in the QDP, suitable Registry Tabs (Fig. 7.11) can be filled to include registry data of every building, including nine sets of data: general data, characteristic data, dimensional data, original project availability, intervention or modification, structural intervention, vulnerability assessment, reference technical code and "other".

At this point, the *User Reporting* can start and the users can send reports and inspections of damages with the support of the APP for smart devices (*Data Acquisition*, Fig. 7.12).

7.11 Data Processing of the Valencian Building

User reported data are processed in the QDP and technicians can check and validate the reports by reading, editing, or deleting the information (*Data Processing*). Once the check is completed, the *Condition Ratings* can be calculated to quantify the damages of every single component, class of component and for the entire building considered. In addition, in this phase, it is possible to customize the level of danger. In particular specific limit values (showed in Fig. 7.13) are identified for every *Condi-*

Fig. 7.12 First photos and reports received in the QDP on the left, APP main page on the right

Cr	*Local slight damage*	*Local medium damage*	*Local serious damage*
	$Cr < 4$	$4 \leq Cr < 6$	$6 \leq Cr \leq 10$
CCr	*Few damages* and limited to a class of components	*Medium damages* on a class of components	*Serious and extended damage*
	$CCr < 2$	$2 \leq CCr < 6$	$6 \leq CCr \leq 10$
BCr	*Building in good condition*	*Medium widespread damages*	*Serious and extended danger* compromising whole building
	$BCr < 3$	$3 \leq BCr < 5$	$5 \leq BCr \leq 10$

Fig. 7.13 Limit values and the corresponding qualitative expression and associate colour

tion Rating by associating the numerical range to a qualitative expression and a corresponding colour.

Note that an immediate alert when one of the *Condition Ratings* exceeds a predetermined threshold is sent by the QDP to the interested stakeholders.

7.12 Data Analysis and Building Diagnostics in the Valencian Coastline

In conclusion, by exploiting the reports, the *Condition Ratings* (evaluated by Eq. 7.2) and the relative threshold values, it is possible to compare buildings' conditions at a large scale to identify the priority of intervention or a necessary prompt intervention on-site (such operation regards the *Data Analysis*). An example of the localization of the "in worst condition building" that needs maintenance priorities is shown in the upper part of Fig. 7.14. In addition, every building can be represented in a summary

Fig. 7.14 Priority of intervention in a coastal city in the Valencian region and examples of the building summary sheets

sheet displaying the *building condition ratings* (evaluated by Eq. 7.4), some examples of which are shown in the bottom part of Fig. 7.14.

References

Adam, J. M., Moreno, J. D., Bonilla, M., & Pellicer, T. M. (2016). Classification of damage to the structures of buildings in towns in coastal areas. *Engineering Failure Analysis, 70*, 212–221.

Angst, U., Elsener, B., Larsen, C. K., & Vennesland, Ø. (2009). Critical chloride content in reinforced concrete—A review. *Cement and Concrete Research, 39*(12), 1122–1138.

Castro, E. D. (1994). Desenvolvimento de metodologia para manutenção de estruturas de concreto armado. *Master thesis*, Brasilia, Brazil.

Jin, Z., Zhao, X., Zhao, T., & Liu, Y. (2016). Corrosion behavior of steel bar and corrosive cracking of concrete induced by magnesium-sulfate-chloride ions. *Journal of Advanced Concrete Technology, 14*(4), 172–182.

Menero, E. M. O., & Garrido, M. D. P. (2011). El litoral turístico valenciano. Intereses y controversias en un territorio tensionado por el residencialismo. *Boletin de la Asociacion de Geografos Espanoles, 2011* (Vol 56, pp. 177–200).

Moradian, M., Chini, M., & Shekarchi, M. (2015). Durability performance of a structure made with high-performance concrete and prefabricated elements in a marine environment. *Journal of Performance of Constructed Facilities, 29*(6), 04014174.

Moreno, J. D., Bonilla, M., Adam, J. M., Borrachero, M. V., & Soriano, L. (2015). Determining corrosion levels in the reinforcement rebars of buildings in coastal areas. A case study in the Mediterranean coastline. *Construction and Building Materials, 100*, 11–21.

Pradhan, B. (2014). Corrosion behavior of steel reinforcement in concrete exposed to composite chloride–sulfate environment. *Construction and Building Materials, 72*, 398–410.

Sangiorgio, V., Uva, G., & Fatiguso, F. (2018a). Optimized AHP to overcome limits in weight calculation: Building performance application. *Journal of Construction Engineering and Management, 144*(2), 04017101.

Sangiorgio, V., Uva, G., & Fatiguso, F. (2018b). User reporting–based semeiotic assessment of existing building stock at the regional scale. *Journal of Performance of Constructed Facilities, 32*(6), 04018079.

Sangiorgio, V., Iacobellis, G., Adam, J. M., Uva, G., & Fatiguso, F. (2018c). User-Reporting Based Decision Support System for Reinforced Concrete Building Monitoring. *In 2018 IEEE International Conference on Systems, Man, and Cybernetics (SMC)* (pp. 2620–2625). IEEE.

Sangiorgio, V., Pantoja, J. C., Varum, H., Uva, G., & Fatiguso, F. (2019a). Structural degradation assessment of RC buildings: Calibration and comparison of semeiotic-based methodology for decision support system. *Journal of Performance of Constructed Facilities, 33*(2), 04018109.

Sangiorgio, V., Uva, G., Fatiguso, F., & Adam, J. M. (2019b). A new index to evaluate exposure and potential damage to RC building structures in coastal areas. *Engineering Failure Analysis, 100*, 439–455.

Tuutti, K. (1982). *Corrosion of steel in concrete.* Swedish Cement and Concrete Research Institute, Stockolm (Vol. 198, 469 p).